Innovative Pasta with High Nutritional and Health Potential

Innovative Pasta with High Nutritional and Health Potential

Editors

Laura Gazza
Francesca Nocente

MDPI • Basel • Beijing • Wuhan • Barcelona • Belgrade • Manchester • Tokyo • Cluj • Tianjin

Editors
Laura Gazza
CREA - Research Centre for Engineering and Agro-Food Processing
Rome
Italy

Francesca Nocente
CREA - Research Centre for Engineering and Agro-Food Processing
Rome
Italy

Editorial Office
MDPI
St. Alban-Anlage 66
4052 Basel, Switzerland

This is a reprint of articles from the Special Issue published online in the open access journal *Foods* (ISSN 2304-8158) (available at: https://www.mdpi.com/journal/foods/special_issues/pasta_nutritional_healthy).

For citation purposes, cite each article independently as indicated on the article page online and as indicated below:

LastName, A.A.; LastName, B.B.; LastName, C.C. Article Title. *Journal Name* **Year**, *Volume Number*, Page Range.

ISBN 978-3-0365-6748-8 (Hbk)
ISBN 978-3-0365-6749-5 (PDF)

Contents

About the Editors

Laura Gazza

Laura Gazza is senior researcher at CREA Research Centre for Engineering and Agro-Food Processing in Rome, Italy, with a Master degree in Biological Sciences and a Ph.D. in Plant Biotechnology. She carries out research activities in the context of national, regional and international projects in collaboration with Universities, research centers and SMEs in the sectors of molecular genetics, biochemistry and technology of cereals (common and durum wheat, einkorn, oats and other minor cereals with a reduced or no gluten index such as tritordeum, triticale, sorghum and teff). She studies and applies innovative processing technologies (decortication, micronization, air classification, pasta making) in pilot plants aimed at obtaining transformed products, mainly pasta, with increased nutritional and health potential also through the reuse of by-products deriving from other food transformation processes. She published more than 50 peer-reviewed scientific papers. She is member of the Editorial Board of the Open Access Journal 'Foods' by MDPI and of 'Associazione Italiana di Scienza e Tecnologia dei Cereali' (AISTEC). Since 2017, she is Professor of Food Biotechnology at University of Rome "La Sapienza".

Francesca Nocente

Francesca Nocente is a researcher at CREA Research Centre for Engineering and Agro-Food Processing in Rome, Italy. She got a Degree in Biological Sciences at University "La Sapienza" of Rome and a Ph.D. in Biochemistry and Technology applied to Food and Nutrition at Campus Bio-Medico University of Rome. She is involved in research activities on cereals within national and international projects. Experienced in wheat diseases and breeding for disease resistance by classical and marker-assisted selection methods. For more than 15 years she has been studying and applying innovative cereal processing technologies useful for obtaining products, mainly pasta, with improved nutritional and healthy properties, also through the reuse of by-products deriving from agro-food processes. The research activity also concerns studies on malting process of major and minor cereals. She has skills on nutritional and technological quality evaluation of the raw materials and of the processed products. Research activity is documented from more than 60 national and international scientific papers. Since 2014, she is tutor of Food technology within the degree course in Food Sciences and Human Nutrition at the Campus Bio-Medico University of Rome, and since 2022 she is Professor of Food technology.

foods

Editorial

Special Issue: Innovative Pasta with High Nutritional and Health Potential

Laura Gazza * and Francesca Nocente

CREA-Research Center for Engineering and Agro-Food Processing, Via Manziana 30, 00189 Rome, Italy
* Correspondence: laura.gazza@crea.gov.it

This editorial summarizes some of the key challenges in the production of novel pasta formulations in order to obtain high nutritional and healthy products. In the last 20 years, the awareness of the effect of nutrition on human health increased the interest of consumers in food with enhanced nutritional characteristics. Consequently, the agro-food sector was prompted towards the research and development of functional foods presenting healthy features that go beyond their nutritional value. This new trend also involved the pasta sector to look for innovative pasta formulations by replacement or enrichment of durum wheat semolina with functional ingredients from plant [1–3] or animal origin [2,4,5], or by the use of alternative raw materials such as gluten-free cereals [4,6,7], minor cereals [3,8,9], pseudocereals and legumes [2]. Indeed, pasta, due to its versatility, low cost, nutritional value and pleasant taste, is largely consumed worldwide, thus representing a suitable carrier to deliver functional molecules exerting human-health-beneficial effects. The use of functional ingredients or alternative raw materials often change the cooking quality and sensory attributes of pasta, resulting in increased cooking loss, decreased firmness, color changes and pasta flavor. Consequently, the strategy of pasta innovation, along with the identification of functional ingredients, required suitable technological process to obtain novel pasta products, able to retain their beneficial properties, keeping, meanwhile, high cooking and sensorial quality to be attractive to consumers [3,6,10,11]. Dried pasta is the symbol of Italian food, and it is certainly no coincidence that 7 out of the 11 articles published in this Special Issue, come from Italian research groups [2,3,6–8,10,11]. Pasta functionalization has been mainly reached by fortification with dietary fiber [3,6], antioxidants [4,5], proteins [2], vitamins and minerals [1], which are present in a limited amount in durum wheat semolina. The consumption of high-fiber pasta contributes to reducing blood pressure, cholesterol levels, risk of colon cancer and coronary heart disease and improving the feeling of satiety. Moreover, the enrichment with dietary fiber also reduces the glycemic index of pasta, a health aspect often considered in this Special Issue [6–8,10,11]. Generally, the presence of dietary fiber results in detrimental effects on the cooking and sensory qualities of pasta, although the source of fiber used and the level of enrichment, as well as the technological process applied, largely influence the overall quality of unconventional pasta products. The technological procedures useful to overcome the negative effect of wheat bran enrichment include micronization [3,6], air classification, debranning, fermentation [10] and enzymatic hydrolysis. Besides the presence of dietary fiber, the importance of wholegrain consumption is also related to the presence of health-promoting antioxidant compounds, mostly present in the outer kernel layers. Polyphenols are the most abundant plant antioxidants and the long-term consumption of foods rich in polyphenols is considered healthier by reducing the risk of some types of cancer, diabetes and neurodegenerative and cardiovascular diseases. Fruits and vegetables are particularly rich in polyphenols as well as vitamins, minerals and fiber. Hence, they have been used as ingredients mainly to enhance the antioxidant activity of pasta by enrichment with flour from dried fruit and vegetables [1–3], with pureed or with antioxidant compounds extracted from them. Recently, pasta functionalization has also been realized by the use

Citation: Gazza, L.; Nocente, F. Special Issue: Innovative Pasta with High Nutritional and Health Potential. *Foods* **2022**, *11*, 2448. https://doi.org/10.3390/foods11162448

Received: 1 August 2022
Accepted: 12 August 2022
Published: 14 August 2022

of novel ingredients such as edible insects, fish [4,5], seaweed [2] and meat as a source of proteins, fiber, vitamins, minerals and beneficial fats. For example, the partial replacement of semolina with cricket flour increase proteins, fiber, unsaturated fatty acids, minerals and antioxidant activity, while it decreases carbohydrate content of enriched pasta. Moreover, the growing demand for gluten-free products for people affected by celiac and non-celiac gluten sensitivity and for consumers who prefer a gluten-free diet has been satisfied by the presence on the market of a broad range of high-quality gluten-free pasta. Although the lack of gluten makes the processability of dough into pasta very difficult, the replacement of semolina with gluten-free cereals (rice, corn, sorghum), pseudocereals (quinoa, chia, buckwheat), and legumes (beans, fava beans, peas, chickpeas, lentils, lupins) and the use of additives (gums, emulsifiers or hydrocolloids) and appropriate technologies has allowed industries to produce gluten-free pasta with acceptable sensorial characteristics [4,6,7].

Finally, the pasta sector does not overlook the transition towards a circular economy, one of the European Green Deal pillars, essential for the sustainability of food systems in terms of reducing both resource consumption and waste into the environment. These issues can be addressed by the re-use of food by-products and food wastes to increase the nutritional value of pasta, being often a good source of proteins, minerals, fatty acids, fiber and bioactive compounds. Four articles of this Special Issue reported the results of the up-cycling of fruit and vegetable by-products [1,3] or fish processing [4,5] by the production of dry pasta, with an increased potential nutritional value, as an example of "circular" innovation in the pasta food chain.

Funding: This research received no external funding.

Conflicts of Interest: The authors declare no conflict of interest.

References

1. Bchir, B.; Karoui, R.; Danthine, S.; Blecker, C.; Besbes, S.; Attia, H. Date, Apple, and Pear By-Products as Functional Ingredients in Pasta: Cooking Quality Attributes and Physicochemical, Rheological, and Sensorial Properties. *Foods* **2022**, *11*, 1393. [CrossRef] [PubMed]
2. Messia, M.C.; Cuomo, F.; Falasca, L.; Trivisonno, M.C.; De Arcangelis, E.; Marconi, E. Nutritional and Technological Quality of High Protein Pasta. *Foods* **2021**, *10*, 589. [CrossRef] [PubMed]
3. Nocente, F.; Natale, C.; Galassi, E.; Taddei, F.; Gazza, L. Using Einkorn and Tritordeum Brewers' Spent Grain to Increase the Nutritional Potential of Durum Wheat Pasta. *Foods* **2021**, *10*, 502. [CrossRef]
4. Aínsa, A.; Vega, A.; Honrado, A.; Marquina, P.; Roncales, P.; Gracia, J.A.B.; Morales, J.B.C. Gluten-Free Pasta Enriched with Fish by-Product for Special Dietary Uses: Technological Quality and Sensory Properties. *Foods* **2021**, *10*, 3049. [CrossRef] [PubMed]
5. Ainsa, A.; Honrado, A.; Marquina, P.L.; Roncalés, P.; Beltrán, J.A.; Calanche, M.J.B. Innovative Development of Pasta with the Addition of Fish By-Products from Two Species. *Foods* **2021**, *10*, 1889. [CrossRef] [PubMed]
6. Taddei, F.; Galassi, E.; Nocente, F.; Gazza, L. Innovative Milling Processes to Improve the Technological and Nutritional Quality of Parboiled Brown Rice Pasta from Contrasting Amylose Content Cultivars. *Foods* **2021**, *10*, 1316. [CrossRef] [PubMed]
7. Cervini, M.; Gruppi, A.; Bassani, A.; Spigno, G.; Giuberti, G. Potential Application of Resistant Starch Sorghum in Gluten-Free Pasta: Nutritional, Structural and Sensory Evaluations. *Foods* **2021**, *10*, 908. [CrossRef] [PubMed]
8. Pandolfo, A.; Messina, B.; Russo, G. Evaluation of Glycemic Index of Six Different Samples of Commercial and Experimental Pasta Differing in Wheat Varieties and Production Processes. *Foods* **2021**, *10*, 2221. [CrossRef] [PubMed]
9. Bobryk-Mamczarz, A.; Kiełtyka-Dadasiewicz, A.; Rachoń, L. Usefulness of Hulled Wheats Grown in Polish Environment for Wholegrain Pasta-Making. *Foods* **2021**, *10*, 458. [CrossRef] [PubMed]
10. Fois, S.; Piu, P.P.; Sanna, M.; Roggio, T.; Catzeddu, P. In Vivo and In Vitro Starch Digestibility of Fresh Pasta Produced Using Semolina-Based or Wholemeal Semolina-Based Liquid Sourdough. *Foods* **2021**, *10*, 2507. [CrossRef] [PubMed]
11. Dodi, R.; Bresciani, L.; Biasini, B.; Cossu, M.; Scazzina, F.; Taddei, F.; D'Egidio, M.G.; Dall'Asta, M.; Martini, D. Traditional and Non-Conventional Pasta-Making Processes: Effect on In Vitro Starch Digestibility. *Foods* **2021**, *10*, 921. [CrossRef] [PubMed]

foods

Article

Usefulness of Hulled Wheats Grown in Polish Environment for Wholegrain Pasta-Making

Aneta Bobryk-Mamczarz [1], Anna Kiełtyka-Dadasiewicz [2,*] and Leszek Rachoń [2]

1 PZZ LUBELLA GMW Sp. z o.o. Sp. k., ul. Wrotkowska 1, 20-469 Lublin, Poland; A.Bobryk-Mamczarz@maspex.com
2 Department of Plant Production Technology and Commodity Science, University of Life Sciences in Lublin, Akademicka 15, 20-950 Lublin, Poland; leszek.rachon@up.lublin.pl
* Correspondence: anna.kieltyka-dadasiewicz@up.lublin.pl; Tel.: +48-81-445-6629

Abstract: The best pasta raw material is durum wheat (*Triticum turgidum* subsp. *durum* (Desf.) Husn.). Recently, old wheat species have also attracted interest. The aim of the study was to evaluate their usefulness for industrial pasta production. The technological characteristics of grains and the organoleptic characteristics of pasta obtained from hulled emmer (*T. turgidum* subsp. *dicoccum*) and spelt (*T. aestivum* ssp. *spelta*) were determined and compared to durum wheat, as a standard pasta raw material, and common wheat (*T. aestivum*). All wheats were grown under identical conditions. The hardness of kernels was assessed using the practical size index, wheat hardness index, torque moment, milling work of 50 g of flour, semolina yield, and starch damage. The technological and nutritional values of semolina, i.e., protein and ash content, wet gluten yield and quality, and falling number, were determined. Moreover, the organoleptic characteristics of cooked pasta were analysed in terms of appearance, colour, taste, smell, and consistency. The milling parameters of emmer were comparable to those of durum wheat; moreover, the content of protein, gluten, and ash was higher in emmer. Spelt was found to be similar to common wheat. Hulled wheats, especially emmer, show good quality parameters and can be an alternative raw material for industrial pasta production.

Keywords: innovative pasta raw materials; durum; emmer; spelt; pasta

Citation: Bobryk-Mamczarz, A.; Kiełtyka-Dadasiewicz, A.; Rachoń, L. Usefulness of Hulled Wheats Grown in Polish Environment for Wholegrain Pasta-Making. *Foods* **2021**, *10*, 458. https://doi.org/10.3390/foods10020458

Academic Editors: Laura Gazza, Francesca Nocente, Talwinder S. Kahlon and Antonella Pasqualone

Received: 16 December 2020
Accepted: 15 February 2021
Published: 19 February 2021

1. Introduction

It is widely believed that the best raw material for pasta production is durum wheat (*Triticum durum* Desf.) due to the hardness and vitreousness of its grains, quality of gluten proteins, high yellow pigment content, as well as its light and thin bran layer [1,2]. In recent years, however, much attention has been given to the old species, i.e., hulled wheats such as spelt, emmer, or einkorn, to make traditional products. Hulled wheats are characterised by a lower yield but many of them have a high nutritional value, including higher protein content and wet gluten yield as well as higher content of macro-, micronutrients, and vitamins [3–8]. Regardless of the species, wheat kernels and the resulting semolina should meet certain requirements of pasta producers. The content and quality of protein and gluten, kernel vitreousness and hardness, yellow pigment content, and falling number are relevant. During the production of pasta, the granulation of semolina, its ash content, and the degree of starch damage are also of importance [9]. Pasta producers from different countries have similar requirements for semolina dedicated to the production of pasta. Polish manufacturers of pasta often use the quality guidelines for wheat grain and semolina contained in the non-obligatory Polish Standards [10,11], where, in the case of semolina, recommendations are total ash content max. 0.9%; wet gluten yield min. 30% and its deliquescence up to 13 mm; falling number, being a measure of the amylolytic activity of durum wheat grain, at a minimum level of 300 s; and bulk density of grain at least 75 kg hl^{-1}. Italian law indicates protein content as min. 10.5% and max. 0.9% ash content [12]. On the other hand, Sieber [13], in turn, states that pasta producers in Germany require durum wheat to contain more than 14% protein, grain vitreousness above 75%, parameter b* above 22, and falling number above 220 s.

According to the legislation of France, Greece, and Italy, pasta for local markets can be produced only from durum wheat, without any admixtures of common wheat. In countries such as the USA, Canada, Australia, or Spain, only durum wheat is used by choice in the production of pasta [1,7,14]. In other European countries, pasta producers also use other available raw materials, including high-quality cultivars of common wheat. Other types of wheat may also potentially be the pasta raw material [15,16], but insufficient data are currently available on the use of hulled wheat as a substitute for durum and common wheat in industrial pasta making and their quality [6].

Durum wheat has specific climatic requirements. It grows best in dry, hot, continental climates; therefore, it is mainly grown in the Mediterranean basin, North America, and Kazakhstan. For several years now, it has also been successfully cultivated in Central Europe. Attempts have also been made to obtain durum wheat cultivars for cultivation under temperate conditions [17–20]. The choice of raw materials for the pasta industry is associated not only with their quality, but also their supply. This can be problematic in cases of less yielding and less widespread hulled wheats; climatic constraints on durum cultivation should also be considered [21]. The cultivation of common wheat (*Triticum aestivum* ssp. *vulgare* L.) is the most widespread in the world [22]. Some cultivars of common wheat are of high technological value and are used not only as a bakery raw material, but also for pasta production in some parts of Europe and the world [22–25]. The aim of the present study was to determine the usefulness of two species of hulled wheats, emmer and spelt, for the pasta industry, as compared to two cultivars of durum wheat grown in Poland under temperate conditions, with the indication that 'SMH87' is a local cultivar. [26].

2. Material and Methods

2.1. Plant Material

In order to assess the suitability of the hulled wheats (spelt and emmer) for pasta production, the technological characteristics of kernels were determined and compared with durum wheat, considered the best pasta raw material and common wheat, and now also widely used in the pasta industry (Table 1).

Table 1. Wheat species and cultivars tested.

Wheat Species	Botanical Latin Name	Cultivar
spelt	*Triticum aestivum* ssp. *spelta* (L.) Thell.	'Wirtas'
emmer	*Triticumturgidum* subsp. *dicoccum* (Schrank ex. Schübl.) Thell.	'Bondka'
durum	*Triticum turgidum* subsp. *durum* (Desf.) Husn.	1. 'Floradur' 2. 'SMH87'
common	*Triticum aestivum* L.	'Torridon'

To eliminate the effects of the changeable environmental conditions of raw material production on the quality and usability of the kernels, a field experiment was conducted, in which the same agrotechnical, soil, and climatic conditions were maintained for each cultivar tested, as previously described [21]. The experiment was continued for three consecutive years (2015–2017) in the locality of Hopkie (50°30′28″ N 23°39′40″ E; 221 m a.s.l., Lubelskie Province, Poland) on rendzina soil. The experiment was set up using a random block design, with three replicates, on plots with area of 0.17 ha. The conducted soil tillage was typical of a conventional tillage system and the fertilisation and cultivation of the plantations were used as modern agriculture and managed in accordance with Good Agriculture Practice. The suitability of kernels of individual cultivars for pasta production was assessed annually. After harvesting, the kernels were cleaned and dried to a constant humidity of 13%.

2.2. *Kernel Hardness Characteristics*

The milling value of kernels of each cultivar was evaluated using the following hardness characteristics:

- The particle size index (PSI) expressed as % of the flour produced under the standard grain milling conditions obtained using a Quadrumat Junior mill (Duisburg, Germany). The grain was ground in a mill with grinding gaps of I–II at 0.8 mm; II–III at 0.3 mm; III–IV at 0.1 mm and roll grooves of I and II at 5R/cm; III and IV at 8 R/cm. The obtained grist was sieved in a laboratory sifter on a sieve with a mesh size of 500 μm, so that we received wheat bran and unpurified semolina. Purified semolina (extract 50% ± 2% in relation to the grain) was obtained on the principle of self-sorting and was undersown on sieves wrapped with gauze at 0.8, 0.5, and 0.35 mm. Higher PSI values correspond to the grains of lower hardness. Using a Brabender hardness tester (Nossen, Germany) determined:
- The torque value expressed as the maximum height of the graph in Brabender units (BU).
- The milling work required for fragmentation of 50 g of the grain sample, read as a function of the surface plotted by the recorder.
- The amount of flour produced with a particle size of <120 μm on the laboratory sifter (%).
- The wheat hardness index (WHI) expressed as the ratio of torque in BUs to the quantity of flour (%),
- The yield of non-purified semolina (%) using a Quadrumat Junior mill (Duisburg, Germany).

2.3. *Quality of Wheat Semolina*

From the wholegrain samples, semolina was separated reaching—for all of the tested wheat species—a constant yield of 50% in relation to the grain. The usefulness of semolina was evaluated by determining:

- The total protein content (%) using the Kjeldahl method according to PN-EN ISO 20483 [27].
- The yield of wet gluten (%), its elasticity, and deliquescence according to PN-77/A-74041 [28]. To 50 g of semolina, 25 cm^3 of tap water at 20 °C was added and the dough was kneaded. The dough was rolled into a ball by hand and placed in a steamer for 20 min. After this time, the dough ball was kneaded under running water until the starch was completely washed out (until the water showed no reaction to the presence of starch with Lugol's solution). The obtained gluten ball was pressed by hand to remove excess water, and its weight was determined on a laboratory balance with an accuracy of 0.01 g. The gluten content was converted into 100g of semolina. In order to determine the elasticity, 5 g of the washed gluten was weighed with an accuracy of 0.01 g and formed into a 2 cm long roll. The roller was taken in two hands with the tips of the fingers and brought closer to the millimetre scale so that the lower end of the roller fell to the zero point of the scale in the upper part of the measure. Then, with the fingers of one hand, it was pulled down slightly to the 5 cm point, then the lower end of the roller was released and the behaviour of the pulled-out gluten roller was observed. Gluten elasticity is defined in degrees: 1st degree—elastic gluten, showing the ability to stretch up to 5 cm and return to the zero point of the scale; 2nd degree—moderately elastic gluten, showing the ability to stretch up to 5 cm and return only to half the length, i.e., up to 2.5 cm; 3rd degree—inelastic gluten, showing the ability to stretch, but not shrinking completely, sagging and showing the ability to stretch further; 4th degree—inelastic (short) gluten, breaks before stretching up to a length of 5 cm. Gluten deliquescence was determined as follows: 5 g of gluten was balled and placed on a glass plate with a millimetre mark underneath. The ball diameter was measured in two perpendicular directions. The plate was covered with a glass beaker and placed in an oven at 30 ° C for 60 min. After this time, the ball diameter was

measured again. Gluten deliquescence is expressed in mm as the difference between the final and initial ball diameters.

- The falling number applying the Hagberg–Perten method according to PN-ISO 3093 [29].
- The degree of starch damage (%) using SD Matic (Chopin, Villeneuve-la-Garenne, France), according to AACC 76-31 [30].
- The total ash content (%) according to PN-ISO 2171 [31].
- Colour using a CR-410 Chroma Meter (Konica-Minolta, Tokyo, Japan) in the CIE L*a*b* system, where L* is a measure of lightness (ranging from 0 for ideal black to 100 for ideal white); a*, where negative values indicate green and positive values indicate red; and b*, where positive values indicate yellow. While interpreting the numerical values in the CIE L*a*b* system, it should be assumed that the higher the b* value, the more yellow the sample and the higher the L* value, the lighter the sample. In the paper, we presented only the values of b*, reflecting the yellow colour of the sample, in correlation with L*, responsible for the lightness of the sample. The third element of chromaticity in the CIE L*a*b* system, i.e., a*, which determines the intensity of the red colour, was neglected as its values were around zero (0.13–1.95) and the parameter itself is less important for the quality of pasta.

2.4. Preparation and Organoleptic Evaluation of Pasta

The wholegrain pastas were prepared under repeatable laboratory conditions from each of the wheat cultivars tested to determine their organoleptic characteristics. The following procedures were observed: 80 mL of water at room temperature was slowly added to 200 g of wholegrain wheat semolina obtained after the one-step 20-s milling (Thermomix Vorverk, Wollerau, Switzerland), thus a characteristic pasta dough in the form of a crumble with a moisture content of about 38% (deficient in water) was obtained. The pasta dough was formed in a steel kneading-trough using a rotating agitator for 3 min, embossed through a matrix, and formed into a rotini shape, which was then cut with a slidable knife into four-centimetre pieces (Figure 1). The products were pre-dried at 35 °C for 30 min and then dried in a food dryer at 60 °C (±2 °C) for 6 h to a standard humidity of 12% (Figure 2).

Figure 1. Pasta preparation (photo by A. Bobryk-Mamczarz).

The pasta of each sample was cooked separately in slightly salted water (14 g of NaCl for 2 l of water according to PN-93/A-74130 [32]) over the prescribed minimum cooking time (5 min.), after which the cooked pasta is *al dente* and ready to eat. The organoleptic characteristics were evaluated by a team of five professional certified sensory experts with confirmed sensory sensitivity, professionally involved in organoleptic analysis. The 100-g samples of pasta obtained from each of the cultivars tested were assessed in terms of appearance, colour, taste, smell, and consistency in accordance with PN-87/A-74131 [33].

Each feature was scored on a scale of 1 to 5. The arithmetic means of five evaluations gave the final result for the pasta of each wheat.

Figure 2. Wholegrain semolina and pasta of individual cultivars (photo by A. Bobryk-Mamczarz).

2.5. Statistical Analysis

The results were statistically analysed using the analysis of variance (ANOVA) and Statistica 12 PL software, assessing with Tukey's post hoc HSD (honest significant difference) test; $p \leq 0.05$ was considered statistically significant.

3. Results and Discussion

3.1. Grain Milling Value

The technological suitability of wheat for pasta production can be assessed based on the milling value of grains and the physicochemical parameters of semolina. The highest torque value, corresponding to the wheat of harder grains, was found for the emmer wheat (372 BU) followed by 'Floradur' (358 BU) and 'SMH87' (345 BU) (Table 2). The biggest work (1201 J) required for fragmentation of 50 g of a grain sample was found for 'Floradur', followed by 'SMH87'—1155 J; the results differed significantly. The work needed for fragmentation of the emmer wheat was 9.6% lower, as compared to 'Floradur'. In the case of soft endosperm wheats, i.e., spelt and common wheats, both parameters mentioned above were lower, thus more fine flour < 120 μm was produced during milling. Both wheat cultivars tested (durum and emmer) were also characterised by higher yields of wholegrain semolina, as compared to spelt and common wheat. The above results were consistently confirmed by the WHI and PSI. The highest WHI was recorded for durum wheat cultivars: 'Floradur'—160; 'SMH87'—143; emmer—131. The PSI, which expresses the percentage of flour produced during fragmentation, was the lowest one for wheats with harder endosperms. According to Rachoń [34], the grain of higher hardness wheats is best for pasta production, due to its higher ability to form semolina, the granulation and roughness of milling products, and the amount of fine flour produced during milling. The author has reported a higher torque value, lower amounts of fine flour, a higher WHI, and a lower PSI for hard durum wheat cultivars and lines compared to common wheat 'Sigma'. Moreover, Cacak-Pietrzak and Gondek [35] as well as Wójtowicz et al. [24] have demonstrated a higher hardness of common wheat grains, as compared to spelt, which is consistent with our results.

Table 2. Kernel hardness characteristics (means of the years 2015–2017).

Wheat Species	Torque Value [BU]	Milling Work of 50 g of Flour [J]	Amount of Flour with a Particle Size <120 μm [%]	wholegrain Semolina Yield [%]	WHI	PSI
spelt	252 E*	759 E*	5.64 B*	73.4 D*	49 D*	16.3 A*
emmer	372 A	1086 C	2.89 C	76.7 A	131 C	7.2 C
durum 1	358 B	1201 A	2.34 D	76.4 B	160 A	7.2 C
durum 2	345 C	1155 B	2.38 D	76.5 AB	143 B	7.0 C
common	335 D	966 D	7.04 A	75.0 C	50 D	12.9 B

* Values denoted with the same letter are not statistically significantly different ($p \leq 0.05$).

Furthermore, the degree of starch damage in semolina depends on the hardness of kernels. According to Dziki et al. [36], the flour obtained from harder grains is characterised by a higher degree of starch damage, as compared to the soft wheat flour, which is confirmed by our results (Figure 3). For both sifting granulations of purified semolina (through a sieve of 95 and 180 μm), the highest degree of starch damage was found in 'Floradur' and 'SMH87', followed by emmer (8.0 and 5.8%, 7.9% and 5.6%, and 7.1 and 4.4%, respectively). Lower starch damage was observed in the common wheat and the lowest one in the spelt (6.0 and 4.1% and 4.0 and 2.6%, respectively).

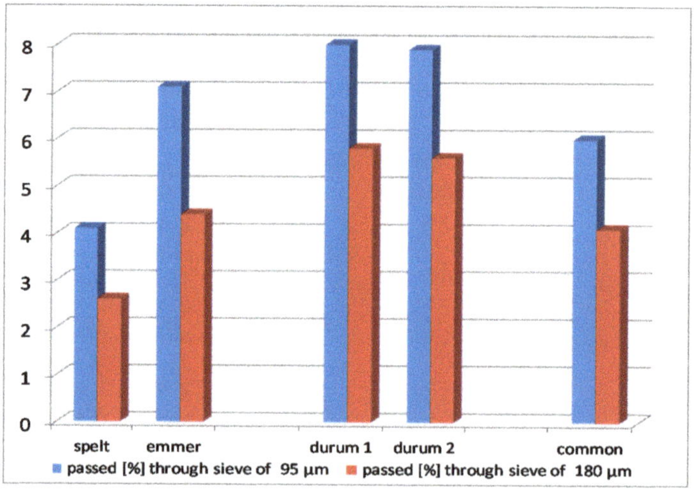

Figure 3. Starch damage degree.

Milling conditions have a significantly greater impact on the degree of starch damage than the choice of wheat cultivars [37,38]. The degree of starch damage is a determinant of milling quality assessment, which may cause the water absorption of semolina during kneading the dough to be too high, and during pasta cooking, too much water-soluble amylose is released into the solution. According to Szafrańska [39], the damaged starch absorbs more water during dough formation, as compared to the undamaged starch; hence, less water is left for proper gluten network development. During the milling of durum wheat into semolina for the production of pasta, the aim is to minimise starch damage. The optimal degree of starch damage also depends on the amount of total protein [39]. Therefore, in the case of wheat milling for pasta production, the degree of starch damage should be as low as possible. Our results indicated that the degree of starch damage increased with an increase in granulation for all the samples tested. The highest degree of starch damage (in the material of the same granulation) was recorded in the case of both durum and emmer wheat cultivars, and the lowest in spelt. Much lower starch damage in products of the same species and cultivars, but with higher granulation, recommends such grinding for pasta purposes, which results in as little flour as possible and as much grist as possible.

3.2. Quality Parameters of Semolina

3.2.1. Protein Content and Gluten Yield

The protein content in the raw material is one of the most important parameters determining its suitability for pasta production [1,34,40]. The highest total protein content in purified semolina was found for emmer—18.0%; significantly lower values were determined in spelt semolina—14.8%; and in both durum wheat cultivars—'SMH87' and 'Floradur'—4.1 and 4.3 percentage points (p.p.), respectively, as compared to emmer (Table 3). The lowest total protein content was found in common wheat semolina—12.2%. The highest amount of gluten was washed out from hulled wheat semolina: in emmer, the percentage of gluten was 38.0%, a significantly lower amount (by 4.6 p.p.) was found in spelt. The amount of gluten washed out from wheats of both durum cultivars was comparable (28.7–28.8%); the lowest yield of gluten was observed for common wheat—24.3%, which was lower by 13.7 p.p. than the yield observed in emmer. Likewise, in the studies by Branković et al. [41], Woźniak [42], Geisslitz et al. [43], and Rachoń [20], the durum wheat contained more total protein and gluten than the common wheat. Majewska et al. [44] reported a higher content of wet gluten in flours from seven spelt cultivars and a higher total protein content (except for one cultivar—Celario), as compared to common wheat flour. In contrast, Sobczyk et al. [45] reported a lower content of protein and gluten proteins in spelt than in common wheat; in turn, Frakolaki [46] reports that spelt had more protein but less gluten than common wheat. According to Suchowilska et al. [47], the total protein content was higher in emmer than in spelt, which was confirmed in our study.

Table 3. Quality parameters of fine semolina (means of the years 2015–1017).

Wheat Species	Protein Content [%]	Wet Gluten Yield [%]	Deliquescence of Gluten [mm]	Elasticity of Gluten [Degrees]	Falling Number [s]	Total Ash Content [%]
spelt	14.8 B*	33.4 B*	4.8 BC*	II	388 D*	0.70 D*
emmer	18.0 A	38.0 A	13.0 A	III	452 C	1.27 A
durum 1	13.7 C	28.7 C	4.5 C	II	506 A	0.80 C
durum 2	13.9 C	28.8 C	5.6 B	II	477 B	0.85 B
common	12.2 D	24.3 D	2.8 D	II	375 E	0.63 E

* Values denoted with the same letter are not statistically significantly different ($p \leq 0.05$).

3.2.2. Quality of Gluten

The quality of gluten was assessed by determining the deliquescence and elasticity of gluten. The lowest deliquescence was determined in gluten from the common wheat semolina—2.8 mm; its value was significantly different compared to other species. The deliquescence of gluten from 'Floradur' (4.5 mm) was not significantly different from the value observed for spelt semolina—4.8 mm (Table 3). Furthermore, the deliquescence of gluten from spelt semolina was not significantly different from that observed in 'SMH87'—5.6 mm. The highest deliquescence was found in emmer gluten—13.0 mm. Emmer gluten was characterised by the highest elasticity (III degree). The remaining species were characterised by second-degree elasticity. Rachoń [34] observed deliquescence of 7–13 mm in eight lines and cultivars of durum wheat. The author has emphasised that gluten in the pasta industry cannot be too short and strong (deliquescence should not be too low) or too weak (deliquescence should not be too high as well). According to the study results reported by Rachoń et al. [2], the durum wheat flour was characterised by gluten deliquescence of 6.3–6.6 mm; in the common wheat flour, this value was 1.5 mm while in the spelt flour—4.0–4.4 mm.

3.2.3. Falling Number

The highest falling number was recorded for 'Floradur'—506 s. The falling number for 'SMH87' was significantly lower—477 s. The falling number of semolina obtained from emmer was 452 s, and from spelt—388 s. The lowest falling number was found in common wheat semolina—375 p. The Hagberg falling number indicates the activity of alpha-amylase in the kernel; in the case of sprouted grain, the falling number is expected to be lower than in grains with low alpha-amylase activity [48,49]. According to Woźniak [42], the falling number in durum wheat was higher than 300 s, regardless of the level of agrotechnics. In the study by Sobczyk et al. [45], the falling number in spelt flour was 257–364 s, while in common wheat flour—271 s. Moreover, Majewska et al. [44] have reported a falling number of 215–315 s in spelt flour and of 296 s in ordinary wheat flour. According to Krawczyk et al. [50], the value of this parameter was 270–331 s in spelt flour and 296 s in common wheat flour. Stolickova and Konvalina [51] in their studies on organic farms in the Czech Republic found the lowest falling number in the control sample, which was common wheat—245 s; in emmer, this number was 238–338, and in spelt— 304–356 s. Higher values for falling number in the case of refined semolina in relation to the whole grain may indicate that debranning and germ removal during the milling process reduce the activity of amylolytic enzymes and thus, increase the falling number. Obuchowski [52] states that semolina of the quality indicated for pasta making should have a falling number of 350–450 s. Polish standards [10,11] indicate the value of the falling number in the raw material for pasta production at a minimum of 250 s for common wheat and a minimum of 300 s for durum wheat. Sjoberg et al. [53] distinguish low-quality wheat genotypes as this with falling number below 300 s. The low falling number, in addition to the risk of excessive darkening of the pasta, may affect its stickiness and the formation of clumps due to the process of excessive starch degradation [52]. In our research, the falling number levels in all semolinas met the assumptions for production of over 300 s as recommended above. It can therefore be concluded that all cultivars of the species compared met the requirements for the production of pasta in terms of this parameter.

3.2.4. Ash Content

Statistical analysis showed that the highest total ash content, i.e., residues from burning a sample of grain or flour containing minerals, was observed in emmer semolina—1.27%. A lower score was obtained for 'SMH87'—0.85%; and 'Floradur'—0.80%. Semolina from wheats with soft endosperms was characterised by the lowest mineral content. Similarly, the lowest ash content in common wheat compared to durum, emmer, and spelt was obtained by Geisslitz et al. [43]. The content of ash in spelt wheat semolina was 0.70% and in common wheat semolina—0.63%. Likewise, according to Rachoń [34], the mineral content in durum wheat semolina was higher, as compared to common wheat. The author has indicated that due to the differences in the distribution of mineral compounds throughout the grain (higher content in durum endosperm compared to common wheat), durum wheat semolina contains more ash, compared to wheat flour, which is confirmed by earlier studies [21]. Even though the emmer wheat was also shown to be the richest source of minerals, there was less ash in the entire durum wheat grains than in spelt and ordinary wheat [8,21].

3.2.5. Semolina Colour

The colour of semolina obtained from the wheats tested was found to be characteristic of individual wheat cultivars, while the differences in individual years of the study were slight (Figure 4). The semolina from both durum wheat cultivars studied was characterised by the highest value of b*, which describes the intensity of yellow colour desirable for pasta raw materials. This value fluctuated slightly over the years—the most yellow durum semolina was obtained in 2015. Podolska and Wyzińska [54] have also stressed the differences in the beta-carotene content of durum wheat in the individual years. Furthermore, Fu et al. [55] have observed higher values of b* (27.8–32.7) in Canadian West Amber durum (CWAD) semolina; the parameter depended on several factors, including the granulation of semolina—the finer the semolina, the lower its value at a given pigment content in the raw material. In contrast, Sieber et al. [56], in their study of 46 durum wheat lines collected in Germany, determined lower b* values (in the range 15.0–19.1). Likewise, Subira et al. [9] found b* in the range of 12.9–14.5 for wholegrain durum wheat flours grown under Italian and Spanish conditions, with higher values in modern than in old cultivars. Moreover, comparing the carotenoid content of the three wheat species, Piergiovanni et al. [57] demonstrated the highest carotenoid content in the durum wheat, followed by spelt; the lowest content was observed in emmer. According to Rachoń [34], the yellow pigment content in durum wheat cultivars and lines was 27.7% higher, as compared to the common wheat.

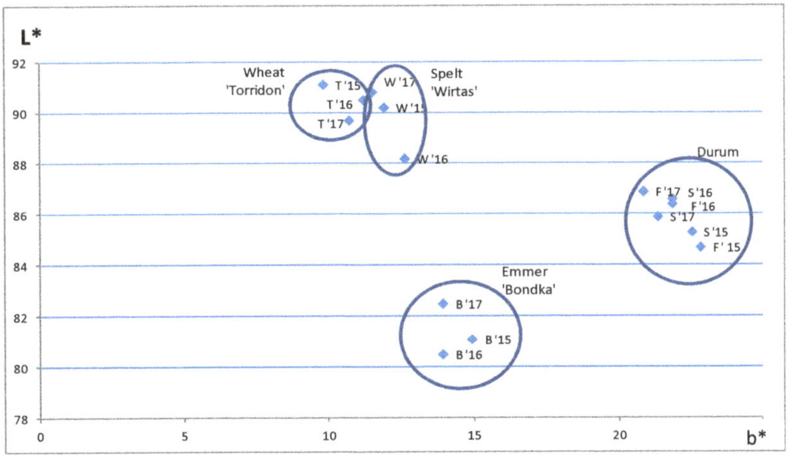

Figure 4. Values of L* responsible for sample lightness and of b* corresponding to yellow colour of purified semolina obtained from the wheats in the individual years of the study (2015–2017) defined in the CIE L*a*b* system. Explanations: W—spelt 'Wirtas'; B—emmer 'Bondka'; S—durum 'SMH87'; F—durum 'Floradur'; T—common wheat 'Torridon'; '15—2015 year; '16—2016 year; '17—2017 year.

In our study, the lowest L*, determining the lightness of the samples, was observed in durum semolina in 2015, which means that these samples were slightly darker than those of 2016–2017. The semolina obtained from the emmer wheat was slightly darker, as compared to the durum semolina (lower L*), while the common wheat and spelt semolina was slightly lighter (higher L*). Fu et al. [55] have reported L* values between 83.8 and 85.5 for the CWAD semolina, i.e., darker than the semolina in our study. Fuad and Prabhasankar [58] have reported the highest L* value (the highest lightness) for common wheat semolina (85.8), followed by durum semolina (81.4); emmer semolina was the darkest one (74.2), which is consistent with our findings. The authors have pointed out that the highest value of L*, i.e., the lightest colour, of the common wheat semolina may be associated with the lowest bran fraction content and lower ash content determined for this wheat species.

Analysis of the b* and L* values of the semolina leads us to expect that the darkest pasta with a shade of yellow will be obtained from emmer wheat. The common wheat and spelt pasta should be expected to be light in colour, although of a low yellow intensity; the durum wheat semolina should provide the most optimal colour (in terms of colour and lightness), regardless of weather conditions during grain maturation.

3.3. Organoleptic Evaluation of Pasta

Pasta obtained from different wheat species differed in sensory characteristics, such as appearance, colour, taste, smell, and consistency (Figure 5). The highest average score (4.4 points) was found for wholegrain pasta from durum wheat 'SMH87': four out of five sensory experts rated it the highest compared to the other pastas. Slightly lower scores were reported for 'Floradur' and spelt pasta—4.2 and 4.1, respectively. Common wheat pasta received the lowest scores in terms of consistency, colour, appearance, and taste.

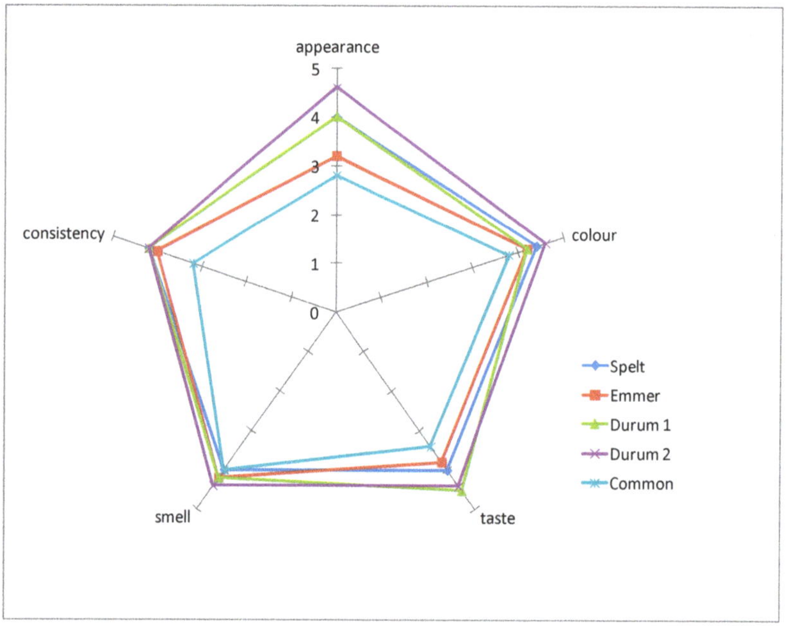

Figure 5. Organoleptic evaluation of pasta produced from individual wheat species.

4. Conclusions

The results regarding grain hardness, semolina technological parameters, and organoleptic evaluation of pasta do not eliminate hulled wheats as a pasta raw material. Hulled wheats are not as good as durum wheat in any quality area, but many parameters, including the entire organoleptic characteristics of pasta, were better than in the common wheat, which is also a proper pasta raw material in some regions of the world. In many features, the differences between emmer and spelt proved important. Emmer had the highest protein, gluten, and ash content of all the cultivars studied. The yield of semolina, kernel hardness, the resulting milling parameters, and the degree of starch damage of the emmer wheat were closest to those of the durum wheat. However, the colour of the semolina and the organoleptic characteristics of the pasta were still weaker than those of durum. In the case of spelt, the parameters in question were similar to those of the common wheat.

In conclusion, the hulled wheats discussed, especially emmer but also spelt, may be considered alternative raw materials for industrial pasta production, provided that their supply is adequate. They are characterised by good quality parameters in the areas studied. These are interesting preliminary results which encourage deepening of the study with a larger number of cultivars and with larger amounts of grain in order to perform trials at the industrial level and also in mixtures with durum wheat.

Author Contributions: Conceptualization, A.B.-M. and A.K.-D.; methodology, A.B.-M. and A.K.-D.; software, A.B.-M.; validation, L.R.; formal analysis, A.B.-M. and A.K.-D.; investigation, L.R.; resources, A.B.-M.; data curation, A.K.-D.; Writing—Original draft preparation, A.K.-D.; Writing—Review and editing, L.R. and A.B.-M.; visualization, A.K.-D.; supervision, L.R.; project administration, A.B.-M.; funding acquisition, L.R. All authors have read and agreed to the published version of the manuscript.

Funding: This research received no external funding.

Institutional Review Board Statement: Not applicable.

Informed Consent Statement: Not applicable.

Data Availability Statement: The data presented in this study are available on request from the corresponding author.

Conflicts of Interest: The authors declare no conflict of interest.

References

1. Sissons, M. Role of durum wheat composition on the quality of pasta and bread. *Food* **2008**, *2*, 75–90.
2. Rachoń, L.; Szumiło, G.; Stankowski, S. Porównanie wybranych wskaźników wartości technologicznej pszenicy zwyczajnej (*Triticum aestivum* ssp. *vulgare*), twardej (*Triticum durum*) i orkiszowej (*Triticum aestivum* ssp. *spelta*). *Fragm. Agron.* **2011**, *28*, 52–59.
3. Arzani, A.; Ashraf, M. Cultivated Ancient Wheats (*Triticum* spp.): A Potential Source of Health-Beneficial Food Products. *Compr. Rev. Food Sci. Food Saf.* **2017**, *16*, 477–488. [CrossRef]
4. Čurná, V.; Lacko-Bartošová, M. Chemical Composition and Nutritional Value of Emmer Wheat (*Triticum dicoccon* Schrank): A Review. *J. Cent. Eur. Agric.* **2017**, *18*, 117–134.
5. Shewry, P.R. Do ancient types of wheat have health benefits compared with modern bread wheat? *J. Cereal Sci.* **2018**, *79*, 469–476. [CrossRef]
6. Kulathunga, J.; Reuhs, B.L.; Simsek, S. A review: Novel trends in hulled wheat processing for value addition. *Trends Food Sci. Technol.* **2020**, *106*, 232–241. [CrossRef]
7. Mefleh, M.; Conte, P.; Fadda, C.; Giunta, F.; Piga, A.; Hassoun, G.; Motzo, R. From ancient to old and modern durum wheat varieties: Interaction among cultivar traits, management, and technological quality. *J. Sci. Food Agric.* **2019**, *99*, 2059–2067. [CrossRef] [PubMed]
8. Rachoń, L.; Bobryk-Mamczarz, A.; Kiełtyka-Dadasiewicz, A. Study on mineral composition of selected wheat species. *Przem. Chem.* **2019**, *98*, 1888–1890. [CrossRef]
9. Subira, J.; Peña, R.J.; Álvaro, F.; Ammar, K.; Ramdani, A.; Royo, C. Breeding progress in the pasta-making quality of durum wheat cultivars released in Italy and Spain during the 20th Century. *Crop Pasture Sci.* **2014**, *65*, 16–26. [CrossRef]
10. The Polish Committee for Standarization. *PKN Polish Standard PN-R-74108:1998 Cereal Grains—Durum Wheat*; PKN: Warszawa, Poland, 1998.
11. The Polish Committee for Standarization. *PKN Polish Standard PN-92-A-74021 Semolina Durum for Alimentary Pasta Production*; PKN: Warszawa, Poland, 1992.
12. Ministero delle Politiche Agricole Alimentari e Forestali. *Decreto del Presidente della Repubblica 9 Febbraio 2001, n.187. Regolamento per la Revisione della Normativa Sulla Produzione e Commercializzazione di Sfarinati e Paste Alimentari, a Norma dell'Articolo 50 della Legge 22 Febbraio 1994, n. 146*; Ministero delle Politiche Agricole Alimentari e Forestali: Rome, Italy.
13. Sieber, A.N. Breeding Winter Durum Wheat for Central Europe: Assessment of Frost Tolerance and Quality on a Phenotypic and Genotypic Level. Ph.D. Thesis, State Plant Breeding Institute University of Hohenheim, Stuttgart, Germany, 30 June 2016. Available online: http://opus.uni-hohenheim.de/volltexte/2016/1219/ (accessed on 26 July 2020).
14. Joubert, M.; Lullien-Pellerin, V.; Morel, M.H. Impact of durum wheat grain composition on semolina yield and pasta quality. In Proceedings of the 15th European Young Cereal Scientists and Technologists Workshop (EYCSTW), Milan/Bergamo, Italy, 26 April 2016.
15. Krawęcka, A.; Sobota, A.; Sykut-Domańska, E. Physicochemical, Sensory, and Cooking Qualities of Pasta Enriched with Oat β-Glucans, Xanthan Gum, and Vital Gluten. *Foods* **2020**, *9*, 1412. [CrossRef]
16. Marconi, E.; Carcea, M.; Graziano, M.; Cubadda, R. Kernel Properties and Pasta Making Quality of Five European Spelt Wheat (*Triticum spelta* L.) Cultivars. *Cereal Chem.* **1999**, *76*, 25–29. [CrossRef]
17. Longin, C.F.H.; Sieber, A.N.; Reif, J.C. Combining frost tolerance, high grain yield and good pasta quality in durum wheat. *Plant Breed.* **2013**, *132*, 353–358. [CrossRef]
18. Polišenská, I.; Jirsa, O.; Spitzer, T.; Sedláčková, I.; Míša, P. Quality and Yield of *Triticum durum* Under Temperate Continental Climate of the Czech Republic. *Acta Univ. Agric. Silvic. Mendel. Brun.* **2018**, *66*, 371–379. [CrossRef]
19. Sieber, A.N.; Würschum, T.; Longin, C.F.H. Vitreosity, its stability and relationship to protein content in durum wheat. *J. Cereal Sci.* **2015**, *61*, 71–77. [CrossRef]

20. Rachoń, L.; Woźniak, A. Variability of spring durum and common wheat yields in the decade 2009–2018 in the Lublin region. *Agron. Sci.* **2020**, *75*, 67–74. [CrossRef]
21. Rachoń, L.; Bobryk-Mamczarz, A.; Kiełtyka-Dadasiewicz, A. Hulled Wheat Productivity and Quality in Modern Agriculture Against Conventional Wheat Species. *Agriculture* **2020**, *10*, 275. [CrossRef]
22. Giraldo, P.; Benavente, E.; Manzano-Agugliaro, F.; Gimenez, E. Worldwide Research Trends on Wheat and Barley: A Bibliometric Comparative Analysis. *Agronomy* **2019**, *9*, 352. [CrossRef]
23. Gońda-Skawińska, M.; Cacak-Pietrzak, G.; Jończyk, K. Estimation of possibility of use of flour from grain of common wheat winter cultivars from organic farming as raw material for pasta production. *Acta Agrophys.* **2020**, *27*, 17–29. [CrossRef]
24. Wójtowicz, A.; Oniszczuk, A.; Kasprzak, K.; Olech, M.; Mitrus, M.; Oniszczuk, T. Chemical composition and selected quality characteristics of new types of precooked wheat and spelt pasta products. *Food Chem.* **2020**, *309*, 125673. [CrossRef]
25. Zarroug, Y.; Mejri, J.; Dhawefi, N.; Ben, S.A.B.; El Felah, M.; Hassouna, M. Comparison of chemical composition of two durum wheat (*Triticum durum* L.) and bread wheat (*Triticum aestivum* L.) germ oils. *Ekin J. Crop Breed. Genet.* **2015**, *1*, 69–73.
26. Banach, J.K.; Majewska, K.; Żuk-Gołaszewska, K. Effect of cultivation system on quality changes in durum wheat grain and flour produced in North-Eastern Europe. *PLoS ONE* **2021**, *16*, e0236617. [CrossRef] [PubMed]
27. The Polish Committee for Standarization. *PKN Polish Standard PN-EN ISO 20483:2013 Cereals and Pulses—Determination of the Nitrogen Content and Calculation of the Crude Protein Content—Kjeldahl Method*; PKN: Warszawa, Poland, 2013.
28. The Polish Committee for Standarization. *PKN Polish Standard PN 77/A-74041 Cereals and Derived Products—Determination of the Gluten Content and Quality*; PKN: Warszawa, Poland, 1977.
29. The Polish Committee for Standarization. *PKN Polish Standard PN-EN ISO 3093:2010 Wheat, Rye and Flours Obtained from Them, Durum Wheat and Semolina—Determination of Falling Number Using the Hagberg—Perten Method*; PKN: Warszawa, Poland, 2010.
30. AACC. *American Association of Cereal Chemists Approved Methods*, 10th ed.; Method 76-31; American Association of Cereal Chemists: St. Paul, MN, USA, 2000.
31. The Polish Committee for Standarization. *PKN Polish Standard PN-EN ISO 2171:2010. Cereals, Pulses and By-Products—Determination of Ash Yield by Incineration*; PKN: Warszawa, Poland, 2010.
32. The Polish Committee for Standarization. *PKN Polish Standard PN 93/A-74130 Pasta—Sampling and Testing Methods*; PKN: Warszawa, Poland, 1993.
33. The Polish Committee for Standarization. *PKN Polish Standard PN-87/A-74131 Pasta*; PKN: Warszawa, Poland, 1987.
34. Rachoń, L. Assessment of usefulness of domestic and foreign cultivars of spring durum wheat (*Triticum durum* Desf.) for pasta production. *Biul. IHAR* **2004**, *231*, 129–137.
35. Cacak-Pietrzak, G.; Gondek, E. Milling value of spelt wheat and wheat grain. *Acta Agrophys.* **2010**, *16*, 263–273.
36. Dziki, D.; Różyło, R.; Laskowski, J. Wheat flour milling and the influence of grain hardness on the process—A review. *Acta Agrophys.* **2011**, *18*, 33–43.
37. Jurga, R. Influence of wheat hardness on the value of milling grain and flour technological value. *Przegl. Zboż. Młyn.* **2010**, *10*, 31–32.
38. Ocieczek, A. Impact of comminution on adsorption properties of gluten-free wheat starch. *Acta Agrophys.* **2013**, *20*, 125–136.
39. Szafrańska, A. Wpływ stopnia uszkodzenia skrobi na wartość technologiczną mąki pszennej (Effect of degree of starch damage on the technological value of flour). *Przegl. Zboż. Młyn.* **2017**, *61*, 18–22.
40. Duda, A.; Adamczak, J.; Chełmińska, P.; Juszkiewicz, J.; Kowalczewski, P. Quality and Nutritional/Textural Properties of Durum Wheat Pasta Enriched with Cricket Powder. *Foods* **2019**, *8*, 46. [CrossRef] [PubMed]
41. Branković, G.; Dodig, D.; Pajić, V.; Kandić, V.; Knežević, D.; Đurić, N.; Živanović, T. Genetic parameters of bread wheat and durum wheat genotypes for technological quality properties in Serbia. *Zemdirb. Agric.* **2018**, *105*, 39–48. [CrossRef]
42. Woźniak, A. The yield and quality of grain of spring wheat (*Triticum aestivum* L.) and hard wheat (*Triticum durum* Desf.) in dependence on agrotechnical level. *Acta Agrophys.* **2006**, *8*, 755–763.
43. Geisslitz, S.; Wieser, H.; Scherf, K.A.; Koechler, P. Gluten protein composition and aggregation properties as predictors for bread volume of common wheat, spelt, durum wheat, emmer and einkorn. *J. Cereal Sci.* **2018**, *83*, 204–212. [CrossRef]
44. Majewska, K.; Dąbkowska, E.; Żuk-Gołaszewska, K.; Tyburski, J. Baking value of flour obtained from the grain of selected spelled (*Triticum spelta* L.) varieties. *Zywn. Nauk. Technol. Ja* **2007**, *14*, 60–71.
45. Sobczyk, A.; Pycia, K.; Jaworska, G. Charakterystyka porównawcza wartości technologicznej ziarna starych odmian i nowych rodów orkiszu (*Triticum Spelta,* L.) oraz ziarna pszenicy zwyczajnej (*Triticum vulgare*). *Zesz. Probl. Post. N. Roln.* **2017**, *589*, 81–91. [CrossRef]
46. Frakolaki, G.; Giannou, V.; Topakas, E.; Tzia, C. Chemical characterization and breadmaking potential of spelt versus wheat flour. *J. Cereal Sci.* **2018**, *79*, 50–56. [CrossRef]
47. Suchowilska, E.; Wiwart, M.; Borejszo, Z.; Packa, D.; Kandler, W.; Krska, R. Discriminant analysis of selected yield components and fatty acid composition of chosen *Triticum monococcum*, *Triticum dicoccum* and *Triticum spelta* accessions. *J. Cereal Sci.* **2009**, *49*, 310–315. [CrossRef]
48. Guan, E.; Yang, Y.; Pang, J.; Zhang, T.; Li, M.; Bian, K. Ultrafine grinding of wheat flour: Effect of flour/starch granule profiles and particle size distribution on falling number and pasting properties. *Food Sci. Nutr.* **2020**, *8*, 2581–2587. [CrossRef] [PubMed]
49. Woźniak, A.; Gontarz, D. The influence of forecrops and different care systems on the quality of grain of spring wheat. *Biulet. IHAR* **2003**, *228*, 33–39.

50. Krawczyk, P.; Leglińska, A.; Kardialik, J. Comparing the technological value of spelt grains to common wheat grains. *Zywn. Nauk. Technol. Ja.* **2008**, *15*, 43–51.
51. Stolickova, M.; Konvalina, P. *Baking Quality of Genetic Resources of Hulled Wheat Species, Grown in Organic Farming*; Mendelnet: Brno, Czech Republic, 2014; pp. 429–434.
52. Obuchowski, W. Ocena jakości surowców zbożowych wykorzystywanych do produkcji makaronu. *Przegl. Zboż. Młyn.* **2008**, *52*, 12–14.
53. Sjoberg, S.M.; Carter, A.H.; Steber, C.M.; Garland-Campbell, K.A. Unraveling complex traits in wheat: Approaches for analyzing genotype × environment interactions in a multienvironment study of falling numbers. *Crop Sci.* **2020**, *60*, 3013–3026. [CrossRef]
54. Podolska, G.; Wyzińska, M. The influence of nitrogen fertilization on selected grain quality parameters of hard wheat (*Triticum durum* Desf.) cv. Komnata. *Fragm. Agron.* **2013**, *30*, 148–158.
55. Fu, B.X.; Wang, K.; Dupuis, B.; Taylor, D.; Nam, S. Kernel vitreousness and protein content: Relationship, interaction and synergistic effects on durum wheat quality. *J. Cereal Sci.* **2018**, *79*, 210–217. [CrossRef]
56. Sieber, A.N.; Würschum, T.; Longin, C.F.H. Evaluation of a semi-controlled test as a selection tool for frost tolerance in durum wheat (*Triticum durum*). *Plant Breed.* **2014**, *133*, 465–469. [CrossRef]
57. Piergiovanni, A.R.; Laghetti, G.; Perrino, P. Characteristics of Meal from Hulled Wheats (*Triticum dicoccon* Schrank and *Triticum spelta* L.): An Evaluation of Selected Accessions. *Cereal Chem.* **1996**, *73*, 732–735.
58. Fuad, T.; Prabhasankar, P. Influences of India's local wheat varieties and additives on quality of pasta. *Food Bioprocess Technol.* **2012**, *5*, 1743–1755. [CrossRef]

foods

MDPI

Article

Using Einkorn and Tritordeum Brewers' Spent Grain to Increase the Nutritional Potential of Durum Wheat Pasta

Francesca Nocente, Chiara Natale, Elena Galassi, Federica Taddei and Laura Gazza *

CREA Research Centre for Engineering and Agro-Food Processing, Via Manziana, 30 00189 Rome, Italy; francesca.nocente@crea.gov.it (F.N.); chiara.natale@tiscali.it (C.N.); elena.galassi@crea.gov.it (E.G.); federica.taddei@crea.gov.it (F.T.)
* Correspondence: laura.gazza@crea.gov.it

Abstract: Brewers' spent grain (BSG), the major by-product of the brewing industry, can be used as a functional ingredient to increase the nutritional value of cereal-based products. In this work, micronized BSG from the einkorn and tritordeum brewing processes were characterized and used to produce four macaroni pasta formulations enriched with BSG at ratios of 5 g and 10 g/100 g of semolina. Einkorn BSG showed the highest values for all the parameters analyzed—proteins, total dietary fiber (TDF) and total antioxidant capacity (TAC)—except for β-glucan. TDF increased up to 42 and 68% in pasta samples enriched with 10% of BSG from tritordeum and einkorn, respectively. The replacement of 10% of semolina with BSG from both cereals significantly increased the β-glucan content and TAC values. Finally, the addition of BSG from einkorn and tritordeum affected to a minimal extent the sensory properties of cooked pasta, which showed higher values of optimal cooking time and cooking loss, but lower total organic matter compared to semolina pasta. Results from the sensorial judgment fell in the good quality ranges for durum wheat pasta; the incorporation of 10% of einkorn BSG resulted in the best compromise in terms of technological, nutritional and sensorial aspects of enriched pasta.

Keywords: brewers' spent grain; dietary fiber; einkorn; functional pasta; tritordeum; upcycling

Citation: Nocente, F.; Natale, C.; Galassi, E.; Taddei, F.; Gazza, L. Using Einkorn and Tritordeum Brewers' Spent Grain to Increase the Nutritional Potential of Durum Wheat Pasta. *Foods* **2021**, *10*, 502. https://doi.org/10.3390/foods10030502

Academic Editors: Joan M. King and Michael Tilley

Received: 30 January 2021
Accepted: 23 February 2021
Published: 26 February 2021

1. Introduction

The agri-food sector presents a major opportunity for the development of a circular economy, since waste and by-products, still rich in carbohydrates, proteins, lipids and complex nutraceuticals, can be managed to realize new valuable products, increasing the sustainability of the food and non-food chains [1–3]. According to the latest data provided by Eurostat, Italy is the leader in the use of recycled materials, with the highest share of circular materials used by the manufacturing system—the average percentage of waste recycling is around 77% of the total amount of waste [4]. Brewers' spent grain (BSG) is the major by-product generated from the beer brewing process and it is estimated that almost 3.4 million tons of BSG are produced annually in the European Union, of which 288,000 tons/year are used in Italy [5,6]. BSG consists of insoluble grain components, mainly cereal seed coat, pericarp and husk, obtained after the extraction of worth. BSG is rich in fiber (30–50% *w*/*w*) and protein (19–30% *w*/*w*), but essential amino acids (e.g., methionine, phenylalanine, tryptophan, histidine and lysine), minerals (e.g., calcium, iron, magnesium, manganese, phosphorus, potassium, sodium), vitamins (e.g., biotin, choline, folic acid, niacin, pantothenic acid and riboflavin) and antioxidant compounds (tocols amd phenolic acids) are also present [5,7]. Variation in its chemical composition can depend on cereal variety or species, location and harvest time, malting and mashing conditions, drying method and the types of adjuncts used during the brewing process [8,9]. To date, the main use of BSG has been as fertilizer or animal feed. However, due to their valuable chemical composition, novel applications of BSG in different areas are expanding, such as extractions of proteins and fiber, sugars and bioactive molecules; energy production; and microorganism cultivation [5,10,11]. The increasing awareness of the relationship between the consumption of healthy foods and well-being led recently to exploiting the nutritional and functional potential of BSG in human nutrition. The use of barley BSG as adjunct in cereal-based food stuffs such as bread, pasta, cookies and snacks can increase their nutritional value, delivering healthier products towards a more sustainable food system [12–16]. Indeed, the incorporation into the human diet of functional compounds present in BSG such as arabinoxylans, β-glucan, tocols and phenolic acids, provides several benefits by contributing to lowering the risk of some diseases, including cancer, gastrointestinal disorders, diabetes, obesity and coronary heart disease [17–20]. Moreover, in recent years, the issue in agricultural biodiversity of reaching low-impact and sustainable agriculture has led to the rediscovery of several minor crops, such as einkorn and tritordeum, which also result in suitable raw material for malting and brewing [21,22]. Einkorn (*Triticum monococcum* L.) is an ancient diploid hulled wheat cultivated until the Bronze Age. In addition to its hardiness, it shows peculiar nutritional features, i.e., high protein content (16–18%), low nitrogen fertilization, high antioxidant content (carotenoids, tocols, conjugated polyphenols, alkylresorcinols and phytosterols), high contents of zinc and iron and a more digestible gluten with respect to the most cultivated wheats [23–26].

Tritordeum (x *Tritordeum* Ascherson and Graebner) is an amphiploid produced by crossing wild barley (*Hordeum chilense* Roem. and Schult) with either tetraploid (*Triticum turgidum* L. ssp. *durum* Desf.) or hexaploid wheat (*Triticum aestivum* L.). These hybridizations produce hexaploid and octoploid tritordeum, respectively [27]. Tritordeum could be a good novel raw material for the production of health-promoting foods, since, especially in the kernel's outmost layers, it is characterized by a high content of fiber (particularly fructans, arabinoxylans and β-glucans to a minor extent), which ranges from 6 to 8%; moreover, it is a source of oleic acid (3–4%) and antioxidant compounds such as phenolic acids and xanthophylls, mainly lutein (4–6 μg/g) esterified with fatty acids, which improve its stability in storage and at high temperatures [28]. In addition, it has been shown to have far fewer gluten immunogenic peptides in comparison with wheat [29].

In this work, BSG recovered from the einkorn and tritordeum brewing processes were investigated as new functional ingredients with which to produce novel formulations of dry pasta with enhanced nutritional potential, which should able to tempt the international pasta market and meet the request of the upcycling of food industry's waste material. Cooking properties, and textural, sensorial and nutritional characteristics of BSG-enriched durum wheat pasta, macaroni shape, were explored, to pursue the dual purposes of improving the sustainability of the brewing sector by valorizing its main process waste, and of producing pasta with enhanced nutritional value.

2. Materials and Methods

2.1. Raw Material

For this study, two different types of BSG were used to enrich pasta. One pilsner-type BSG (BSGE) was derived from a homemade brewing process, performed in a lab-scale plant, of a mix of two malts obtained from a hulled and a dehulled cultivar of einkorn, Norberto and Hammurabi, respectively. The other pilsner-type BSG (BSGT) was derived from the brewing of a malt from a hexaploid tritordeum line.

After the mashing step, BSG were stored at −20 °C and dried at 60 °C for 72 h. The BSG samples kilned to 7% of moisture were micronized at ≤700 μm sieve bya Pulverisette mill (Fritsch, Idar-Oberstein, Germany). Semolina was obtained from a mix of commercial durum wheat varieties grown in Italy during 2019 using the Buhler MLU 202 (Uzwil, Switzerland) plant.

2.2. Pasta-Making Process

Five pasta formulations were produced: S = semolina 100% used as reference; BSGE5 = semolina:BSGE 95:5 (*w:w*); BSGE10 = semolina:BSGE 90:10 (*w:w*); BSGT5 = semolina:BSGT 95:5 (*w:w*); BSGT10 = semolina: BSGT 90:10 (*w:w*) (Figure 1). To obtain proper consistency of the doughs for extrusion, semolina and semolina-BSG formulations were hydrated at different levels depending on the BSG origin. Specifically, the moisture of the dough of semolina pasta and of pasta enriched with BSG from einkorn was 42%, whereas the dough enriched with BSG from tritordeum was 38% water.

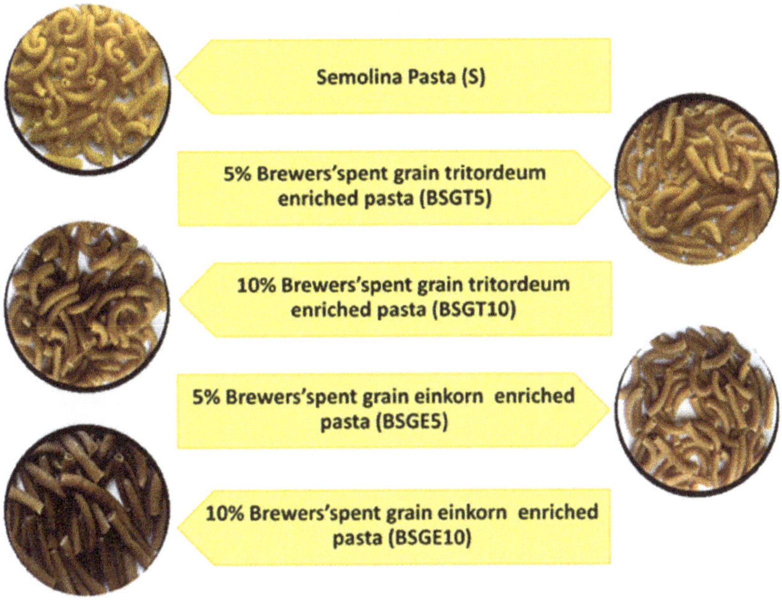

Figure 1. Dry pasta formulations: S = semolina 100%; BSGT5 = semolina:tritordeum Brewers' spent grain (BSGT) 95:5 (*w:w*); BSGT10 = semolina:BSGT 90:10 (*w:w*); BSGE5 = semolina:einkorn BSG 95:5 (*w:w*); BSGE10 = semolina:BSGE 90:10 (*w:w*).

Pasta was obtained using a pilot plant consisting of an extruder (NAMAD, Rome, Italy) with a capacity up to 20 kg/h, equipped with a screw (45 cm in length, 4.5 cm in diameter), which ended with a Teflon-coated die (150 mm diameter) to produce macaroni shape, and of a dryer (AFREM, Lyon, France). Extrusion conditions were those already described by Nocente et al. [12]. For drying, pasta was arranged on frames and dried for 18 h, applying the conditions reported by [12]. The moisture of dried pasta was 12.5%.

2.3. Quality of the Cooked Pasta

Pasta samples were cooked according to the AACC method 16–50 [30]. The optimum cooking time (OCT) of pasta was determined when the white central core of the pasta just disappeared when squeezed between two test glasses, according to D'Egidio et al. [31]. Total organic matter (TOM) of pasta was determined according to D'Egidio et al. [32]. TOM values > 2.1 g/100 g correspond to low quality pasta, between 2.1 and 1.4 g/100 g correspond to good quality pasta, and <1.4 g/100 g correspond to very good quality pasta [33]. Water absorption (WA) was calculated from the weight increase of pasta at the OCT and determined as: $WA = ((w - w_0)/w_0) \times 100$, where w and w_0 were the weights of cooked and raw pasta, respectively [30]. Cooking loss (CL), expressed as grams of matter loss/100 g of raw pasta, was evaluated by weighing the residues of solids lost into the cooking water, after drying overnight at 105 °C. The residue was weighed and reported as percentage [30].

2.4. Sensory Testing

Sensory evaluation was performed, according to D'Egidio et al. [31], by a panel of three trained assessors who evaluated two textural characteristics: stickiness (material adhering to the cooked pasta surface) and firmness (resistance to chewing by the teeth). Each descriptor was scored from 10 to 100. The overall judgment (SJ) was calculated as the arithmetic mean of the scores of each descriptor.

2.5. Basic Composition and Total Antioxidant Capacity of BSG and Pasta Samples

All results are expressed as dry weight (dw) and the moisture content was determined using the thermo balance (Sartorius MA 40, Goettingen, Germany) at 120 °C. All analytical determinations were made in triplicate.

Protein contents of BSG and of BSG-enriched pasta samples were measured by micro-Kjeldhal nitrogen analysis according to ICC 105/2 method [34]. Total dietary fiber (TDF) content was determined using an enzymatic kit for fiber determination (Bioquant, Merck, Darmstadt, Germany) according to the Official Method 991.42 [35]. Ash content was determined according to approved method AACC 08-01.01 [36]. β-glucan content was evaluated by the Megazyme (Bray, Ireland) Mixed-Linkage Beta-Glucan kit [37].

Total antioxidant capacity (TAC) was determined by the "direct method," according to Martini et al. [38].

2.6. Statistical Analysis

Results were expressed as mean ± standard deviation. One-way analysis of variance was performed with MSTATC program (Michigan State University, East Lansing, MI, USA), followed by the Duncan multiple range test for post-hoc comparison of means, applied to assess significant differences ($p < 0.05$) for each considered parameter.

3. Results and Discussion

3.1. Compositional Analysis of Raw Materials

Both spent grains from einkorn (BSGE) and tritordeum (BSGT) showed very high protein contents (Table 1), BSGE having a more than two-fold higher value than the value previously reported for barley BSG [5,12,39]. Likewise, BSGT had higher protein content than the mean value reported by Lynch et al. [5] (20%) for different sources of barley BSG. After all, also einkorn and tritordeum grains showed higher protein contents than barley [27,40]. On the contrary, both BSGE and BSGT's total dietary fiber contents (Table 1) were lower than that reported in previous studies for barley spent grain. This result was expected since BSG from einkorn and tritordeum are devoid of the kernel husk, which contributes most to the higher fiber content recorded in barley BSG [5,12,39]. The β-glucan contents in BSGE and BSGT (Table 1) were lower than that found in barley BSG (2.18%) [12]; nevertheless, it was significantly higher than that registered in durum semolina (0.46%); hence, the addition of einkorn and tritordeum spent grain is supposed to improve the nutritional composition of enriched pasta. The higher values of β-glucan found in BSGT with respect to BSGE (Table 1) reflect the tritordeum genetic origin (barley × wheat). The large amounts of ash in BSG from einkorn and tritordeum (Table 1) represent the greater mineral contents of these cereal grains with respect to wheat and barley kernels [24,41]. The levels of TAC were significantly higher (96%) in BSGE and in BSGT (48%) than in durum semolina (Table 1). TAC values in einkorn were always higher than those found in durum wheat [25], likely due to the synergistic effects of antioxidant compounds such as tocols and carotenoids occurring in higher amounts in *T. monococcum* [23].

Table 1. Basic composition of brewer spent grain (BSG) from einkorn and tritordeum.

SAMPLES	Protein	TDF	β-Glucan	Ash	TAC
	%	%	%	%	mmol TEAC/kg
Einkorn BSGE	32.5 ± 0.1	30.5 ± 0.3	1.00 ± 0.01	3.07 ± 0.02	60.8 ± 0.2
Tritordeum BSGT	21.6 ± 0.3	25.9 ± 0.3	1.660 ± 0.003	2.85 ± 0.02	45.8 ± 0.5

Results are reported as dry weight and expressed as the mean values ± standard deviations for three replications. TDF = total dietary fiber; TAC = total antioxidant capacity; TEAC = trolox equivalent antioxidant capacity.

3.2. Chemical, Technological and Basic Characterizations of Dry Pasta

Though the addition of BSG rich in proteins, fibers and antioxidant compounds to durum semolina is assumed to improve the nutritional potential value of pasta, the percentage of replacement in bakery products is heavily limited by the negative effects of spent grain on the final quality of enriched products. Nevertheless, from our previous findings [12], the incorporation up to 10% of micronized barley BSG resulted in the best compromise in terms of technological, nutritional and sensorial aspects of enriched spaghetti. Then, in the present work, additions of 5 and 10 g of both micronized BSGE and BSGT to 100 g of durum wheat semolina were evaluated in terms of improvement of the nutritional composition of enriched pasta.

The additions of 5 and 10% BSG from einkorn and tritordeum to semolina increased the protein content by 1% on average, with pasta having 10% BSGE showing the highest value (Table 2). The ash content was similar in all pasta samples, even if a gradual increment was observed upon the addition of BSG (Table 2)—it stayed lower than 1.1% though.

Table 2. Basiccomposition of control pasta (S) and brewers' spent grain (BSG) enriched pasta (BSGE5, BSGE10, BSGT5 and BSGT10).

SAMPLES	Protein	Ash	TDF	TAC	β-Glucan
	%	%	%	mmol TEAC/kg	%
S	13.2 ± 0.4 c	0.866 ± 0.008 e	3.4 ± 0.1 d	31 ± 2 c	0.46 ± 0.07 d
BSGE5	14.3 ± 0.2 b	0.963 ± 0.005 c	4.0 ± 0.3 c	34.0 ± 0.7 b	0.50 ± 0.03 cd
BSGE10	15.2 ± 0.3 a	1.098 ± 0.002 a	5.7 ± 0.4 a	37 ± 1 a	0.702 ± 0.001 b
BSGT5	13.8 ± 0.2 bc	0.909 ± 0.008 d	4. 0 ± 0.4 c	33 ± 2 bc	0.56 ± 0.04 c
BSGT 10	14.3 ± 0.4 b	1.000 ± 0.005 b	4.8 ± 0.4 b	34.6 ± 0.6 b	0.81 ± 0.05 a

Results are reported as dry weight and expressed as mean ± standard deviation for 3 replications. Within the same column, values with different letters indicate significant differences determined by Duncan's test ($p < 0.05$). S: semolina 100%; BSGE5: semolina/BSG einkorn (95:5); BSGE10: semolina/BSG einkorn (90:10); BSGT5: semolina/BSG tritordeum (95:5); BSGT10: semolina/BSG tritordeum (90:10). TDF = total dietary fiber; TAC = total antioxidant capacity; TEAC = trolox equivalent antioxidant capacity.

The results showed that TDF content increased according to the magnitude of BSG enrichment. In particular, the amount of TDF was increased by 18% in pasta enriched with 5% BSG from einkorn or tritordeum and by 42% in BSGT10, and even by 68% in BSGE10 (Table 2), with respect to semolina pasta. Noteworthily, accordingly to Regulation (EC) number 1924/2006 [42], BSGE10 pasta could be referred as "rich in fiber," being that it has 6 g dietary fiber/100 g product—the threshold for this nutritional and health claim; however, all pasta samples developed in this study could be labelled as "sources of fiber" due to their having higher than 3 g fiber/100 g. Only 10% BSG resulted in an increase in TAC value higher than the 10% in enriched pasta; lower increments, even if significant, were observed when 5% of semolina was replaced with BSG (Table 2). The addition of BSG to semolina resulted in the increasing of β-glucan content in enriched pasta. In particular, BSGE10 turned out to be increased by 52% with respect to the pasta control, whereas for BSGT10, the result was 74% (Table 2). Additionally, with 5% BSG, we registered an increment of β-glucan content in durum wheat pasta, the increment being more marked when BSG from tritordeum was added (22%) with respect to the addition of BSG from einkorn (9%). According to the health claim by the EFSA [43] related to β-glucan daily intake and its blood LDL cholesterol-lowering effect, the consumption of enriched BSG pasta could contribute to reaching the suggested optimal daily intake (3 g/die) of this bioactive compound.

3.3. Characterization and Sensory Evaluation of Cooked Pasta

Enrichments with 5 and 10% of both tritordeum and einkorn BSG had small but significant effects on TOM values (Table 3), which are associated, in each pasta sample, with very good quality, being lower than 1.4% [33]. Likely, the relative low fiber content brought by BSG did not cause a higher amount of starch to be released when cooking with respect to semolina pasta, since, at low concentrations, fiber might be dispersed into the protein/starch matrix. As confirmation, in our previous study [12], spaghetti enriched with different percentages of barley BSG turned out to be richer in fiber content (+20% on average, with respect to present formulations) and consequently resulted in a gradual increments of TOM values with more BSG. The control semolina pasta showed a lower cooking loss (Table 3) than enriched pasta, and statistically significant differences ($p < 0.05$) were observed amongst all samples, with BSGT5 and BSGT10 showing the highest values. Cooking loss parameter is one of the most important traits that affect consumer acceptance of fiber-enriched pasta. The increase in cooking loss in fiber-richer pasta is presumably due to weakening of protein network by the presence of TDF. Anyway, BSG-enriched pasta showed cooking losses below the values reported for good quality durum wheat pasta (<6.5%) [44].

Table 3. Cooking properties of pasta with different blends of BSG/semolina.

SAMPLES	TOM	CL	OCT	WA
	%	%	min′ sec″	g
S	1.28 ± 0.08 a	4.61 ± 0.04 e	7′30″ ± 5″ c	181.80 ± 0.02 e
BSGE5	1.07 ± 0.04 bc	4.78 ± 0.09 d	8′10″ ± 5″ a	185.80 ± 0.01 c
BSGE10	1.00 ± 0.04 c	4.85 ± 0.01 c	8′10″ ± 5″ a	197.00 ± 0.03 a
BSGT5	1.01 ± 0.09 bc	5.14 ± 0.04 a	7′50″ ± 5″ b	184.20 ± 0.02 d
BSGT10	1.10 ± 0.01 b	4.95 ± 0.02 b	8′00″ ± 5″ ab	193.50 ± 0.01 b

Results are reported as dry weight and expressed as mean value ± standard deviation for 3 replications. Within the same column, values with different letters indicate significant differences determined by Duncan's test ($p < 0.05$). S: semolina 100%; BSGE5: semolina/BSG einkorn (95:5); BSGE10: semolina/BSG einkorn (90:10); BSGT5: semolina/BSG tritordeum (95:5); BSGT10: semolina/BSG tritordeum (90:10). TOM: total organic matter; CL: cooking loss; OCT: optimal cooking time; WA: water absorption.

A significant increase in optimal cooking time (OCT), determined by the disappearance of the starchy core, was observed upon the addition of BSG (Table 3), mainly in samples enriched with einkorn spent grain. An increase in cooking time also caused an increase in water absorption (Table 3), since more water can diffuse within starch and gluten, facilitating starch granules swelling and brokage [45].

The highest global sensorial judgment (Table 4) was found for durum semolina pasta and for pasta with 10% einkorn spent grain substitution; slightly lower scores, in the same class of quality, were reported for BSGT5 and BSGE5 (Table 4). Pasta with 10% BSGT replacement received the lowest scores in terms of both stickiness and firmness. It should be taken into account that the detrimental or positive effects of fiber addition to semolina depend also on the source of the fiber itself, as previously observed [45–47]. In particular, it could be inferred that the addition of einkorn spent grain affected to a minor extent the global quality of enriched pasta, likely due to higher protein content and lower β-glucan content than are found in tritordeum spent grains. Indeed, the inclusion of β-glucan in pasta formulation decreased the firmness and increased the stickiness value [48–50].

Table 4. Sensory evaluation of pasta samples.

SAMPLES	Cooking Quality Parameters		
	Stickiness	Firmness	Global Sensorial Judgement
S	90	75	83
BSGE5	80	75	78
BSGE10	85	80	83
BSGT5	85	70	78
BSGT10	70	70	70

S: semolina 100% (control); BSGE5: semolina-BSG einkorn (95:5); BSGE10: semolina-BSG einkorn (90:10); BSGT5: semolina-BSG tritordeum (95:5); BSGT10: semolina-BSG tritordeum (90:10). For stickiness: 10–20 = very high, 21–40 = high, 41–60 = rare, 61–80 = minimal and 81–100 = absent; for firmness: 10–20 = absent, 21–40 = rare, 41–60 = sufficient, 61–80 = good and 81–100 = very good. For global sensorial judgment, 54 = scarce; 55–64 = sufficient; 65–74 = good; 75–100 = very good.

4. Conclusions

Many recent works have investigated the effects of adding spent barley grains to different cereal products. Currently, the growing interest in cereal alternatives to barley for the malting and brewing processes has made available considerable amounts of brewers' spent grain of different plant origins. The present study indicated that the use of BSG, aimed at increasing the nutritional potential of pasta, should take into consideration the cereal species used in the malting and brewing processes. Indeed, pasta enriched with 10% spent einkorn grain showed the best quality in terms of technological performance, and the nutritional and sensorial parameters investigated. Nevertheless, the findings pointed out that the addition of micronized BSG from both cereals to semolina resulted in pasta with notable increases in protein, TDF and β-glucan content and to a minor extent in TAC levels, along with good sensorial quality. The increasing demand for healthier foods is encouraging the quest for novel raw materials and products. Further exploration should be aimed at considering the consumers' attitude, in terms of flavor and mouthfeel properties, towards these new sorts of functional pasta, such as that proposed in the present paper. Moreover, it would be worth performing a cost–benefit economic analysis of the upcycling of BSG in the functional pasta formulations.

Author Contributions: Conceptualization, F.N. and L.G.; methodology, F.N., F.T. and E.G.; validation, F.N., F.T., E.G. and L.G.; formal analysis, F.N. and E.G.; investigation, C.N., F.T. and E.G.; resources, L.G.; data curation, F.N. and L.G.; writing—original draft preparation, F.N. and L.G.; writing—review and editing, F.N., F.T., E.G. and L.G.; visualization, C.N. and F.T.; supervision, L.G.; project administration, L.G.; funding acquisition, L.G. All authors have read and agreed to the published version of the manuscript.

Funding: This research was funded by Regione Marche, Grant n. B78I18001060005; Regione Siciliana, Grant n. G64I20000520009 and "The APC was funded by Royalties from Einkorn breeders' rights".

Institutional Review Board Statement: Not applicable.

Informed Consent Statement: Not applicable.

Data Availability Statement: Data is contained within the article.

Acknowledgments: The Authors would like to thank Roberto Mortaro, Ester Gosparini and Alessandra Arcangeli of CREA-IT of Rome for their technical assistance in pasta making process and sensorial judgment.

Conflicts of Interest: The authors declare no conflict of interest. The funders had no role in the design of the study; in the collection, analyses, or interpretation of data; in the writing of the manuscript, or in the decision to publish the results.

References

1. Socaci, S.A.; Fărcaş, A.C.; Vodnar, D.C.; Tofană, M. Food Wastes as Valuable Sources of Bioactive Molecules. In *Superfood and Functional Food-The Development of Superfoods and Their Roles as Medicine*; Shiomi, N., Waisundara, V., Eds.; IntechOpen: London, UK, 2017.
2. Muscio, A.; Sisto, R. Are Agri-Food Systems Really Switching to a Circular Economy Model? Implications for European Research and Innovation Policy. *Sustainability* **2020**, *12*, 5554. [CrossRef]
3. Tlais, A.Z.A.; Fiorino, G.M.; Polo, A.; Filannino, P.; Di Cagno, R. High-Value Compounds in Fruit, Vegetable and Cereal Byproducts: An Overview of Potential Sustainable Reuse and Exploitation. *Molecules* **2020**, *25*, 2987. [CrossRef] [PubMed]
4. Presentation Report Study Publishing Organisation or Company: Circular Economy Network. March 2020. Available online: https://circulareconomynetwork.it/rapporto-economia-circolare-2020/ (accessed on 3 December 2020).
5. Lynch, K.M.; Steffen, E.J.; Arendt, E.K. Brewers' spent grain: A review with an emphasis on food and health. *J. Inst. Brew.* **2016**, *122*, 553–568. [CrossRef]
6. Assobirra. Annual Report for the year 2017. 2018. Available online: http://www.assobirra.it/ (accessed on 1 December 2020).
7. Ikram, S.; Huang, L.; Zhang, H.; Wang, J.; Yin, M. Composition and nutrient value proposition of brewers spent grain. *J. Food Sci.* **2017**, *82*, 2232–2242. [CrossRef] [PubMed]
8. Santos, M.; Jiménez, J.J.; Bartolomé, B.; Gómez-Cordovés, C.; del Nozal, M.J. Variability of brewer's spent grain within a brewery. *Food Chem.* **2003**, *80*, 17–21. [CrossRef]
9. Singh, A.P.; Mandal, R.; Shojaei, M.; Singh, A.; Kowalczewski, P.L.; Ligaj, M.; Pawlicz, J.; Jarzebski, M. Novel drying methods for sustainable upcycling of brewers' spent grains as a plant protein source. *Sustainability* **2020**, *12*, 3660. [CrossRef]
10. Chetrariu, A.; Dabija, A. Brewer's Spent Grains: Possibilities of Valorization, a review. *Appl. Sci.* **2020**, *10*, 5619. [CrossRef]
11. Mussatto, S.I. Brewer's spent grain: A valuable feedstock for industrial application. *J. Sci. Food Agric.* **2014**, *94*, 1264–1275. [CrossRef]
12. Nocente, F.; Taddei, F.; Galassi, E.; Gazza, L. Upcycling of brewers' spent grain by production of dry pasta with higher nutritional potential. *LWT Food Sci. Technol.* **2019**, *114*, 108421. [CrossRef]
13. Cappa, C.; Alamprese, C. Brewer's spent grain valorization in fiber-enriched fresh egg pasta production: Modelling and optimization study. *LWT Food Sci. Technol.* **2017**, *82*, 464–470. [CrossRef]
14. Farcas, A.; Tofana, M.; Socaci, S.; Mudura, E.; Scrob, S.; Salanta, L.; Muresan, V.J. Brewers' spent grain—A new potential ingredient for functional foods. *J. Agroaliment. Process. Technol.* **2014**, *20*, 137–141.
15. Ozturk, S.; Ozboy, O.; Cavidoglu, I.; Koksel, H. Effects of brewers' spent grain on the quality and dietary fibre content of cookies. *J. Inst. Brew.* **2002**, *108*, 23–27. [CrossRef]
16. Amoriello, T.; Mellara, F.; Galli, V.; Amoriello, M.; Ciccoritti, R. Technological properties and consumer acceptability of bakery products enriched with brewers'spent grains. *Foods* **2020**, *9*, 1492. [CrossRef] [PubMed]
17. Fastnaught, C.E. Barley fibre. In *Handbook of Dietary Fibre*; Cho, S., Dreher, M., Eds.; Marcel Dekker: New York, NY, USA, 2001.
18. McCarthy, A.L.; O'Callaghan, Y.C.; Piggott, C.O.; FitzGerald, R.J.; O'Brien, N.M. Brewers' spent grain; bioactivity of phenolic component, its role in animal nutrition and potential for incorporation in functional foods: A review. *Proc. Nutr. Soc.* **2013**, *72*, 117–125. [CrossRef]
19. Steiner, J.; Procopio, S.; Becker, T. Brewer's spent grain: Source of value-added polysaccharides for the food industry in reference to the health claims. *Eur. Food Res. Technol.* **2015**, *241*, 303–315. [CrossRef]
20. Reis, S.F.; Abu-Ghannam, N. Antioxidant capacity, arabinoxylans content and in vitro glycaemic index of cereal-based snacks incorporated with brewer's spent grain. *LWT Food Sci. Technol.* **2014**, *55*, 269–277. [CrossRef]
21. Mayer, H.; Marconi, O.; Perretti, G.; Sensidoni, M.; Fantozzi, P. Investigation of the suitability of hulled wheats for malting and brewing. *J. Am. Soc. Brew. Chem.* **2011**, *69*, 116–120. [CrossRef]
22. Zdaniewicz, M.; Pater, A.; Hrabia, O.; Duliński, R.; Cioch-Skoneczny, M. Tritordeum malt: An innovative raw material for beer production. *J. Cereal Sci.* **2020**, *96*, 103095. [CrossRef]
23. Hidalgo, A.; Brandolini, A.; Pompei, C.; Piscozzi, R. Carotenoids and tocols of einkorn wheat (Triticum monococcum ssp. monococcum L.). *J. Cereal Sci.* **2006**, *44*, 182–193. [CrossRef]
24. Hidalgo, A.; Brandolini, A. Nutritional properties of einkorn wheat (*Triticum monococcum* L.). *J. Sci. Food Agric.* **2014**, *94*, 601–612. [CrossRef]
25. Ciccoritti, R.; Taddei, F.; Gazza, L.; Nocente, F. Influence of kernel thermal pre-treatments on 5-n-alkylresorcinols, polyphenols and antioxidant activity of durum and einkorn wheat. *Eur. Food Res. Technol.* **2020**. [CrossRef]
26. Di Stasio, L.; Picascia, S.; Auricchio, R.; Vitale, S.; Gazza, L.; Picariello, G.; Gianfrani, C.; Mamone, G. Comparative Analysis of in vitro Digestibility and Immunogenicity of Gliadin Proteins From Durum and Einkorn Wheat. *Front. Nutr.* **2020**, *7*, 56. [CrossRef] [PubMed]
27. Martın, A.; Alvarez, J.B.; Martın, L.M.; Barro, F.; Ballesteros, J. The development of tritordeum: A novel cereal for food processing. *J. Cereal Sci.* **1999**, *30*, 85–95. [CrossRef]
28. Giordano, D.; Reyneri, A.; Locatelli, M.; Coïsson, J.D.; Blandino, M. Distribution of bioactive compounds in pearled fractions of tritordeum. *Food Chem.* **2019**, *301*, 125228. [CrossRef]

29. Vaquero, L.; Comino, I.; Vivas, S.; Rodríguez-Martín, L.; Giménez, M.J.; Pastor, J.; Sousa, C.; Barro, F. Tritordeum: A novel cereal for food processing with good acceptability and significant reduction in gluten immunogenic peptides in comparison with wheat. *J. Sci. Food Agric.* **2018**, *98*, 2201–2209. [CrossRef]
30. American Association of Cereal Chemists. Pasta and noodle cooking quality-firmness. In *Approved Methods of Analysis, 66-50.01*; American Association of Cereal Chemists: St. Paul, MN, USA, 2000.
31. D'Egidio, M.G.; Mariani, B.M.; Nardi, S.; Novaro, P.; Cubadda, R. Chemical and technological variables and their relationships: A predictive equation for pasta cooking quality. *Cereal Chem.* **1990**, *67*, 275–281.
32. D'Egidio, M.G.; De Stefanis, E.; Fortini, S.; Galterio, G.; Nardi, S.; Sgrulletta, D.; Bozzini, A. Standardization of cooking quality analysis in macaroni and pasta products. *Cereal Foods World* **1982**, *27*, 367–368.
33. D'Egidio, M.G.; Nardi, S. Textural measurement of cooked spaghetti. In *Pasta and Noodle Technology*; Kruger, J.E., Matsuo, R.R., Dick, J.W., Eds.; AACC: St. Paul, MN, USA, 1996.
34. International Association for Cereal Science and Technology. *ICC Standard Methods (Methods No. 105/2)*; ICC: Vienna, Austria, 2003.
35. Association of Official Analytical Chemists. *Official Methods of Analysis 991*, 16th ed.; Cunniff, P., Ed.; AOAC: Gaithersburg, MD, USA, 1995; p. 42.
36. American Association of Cereal Chemists. *Approved Methods of Analysis, 08-01.01 Ash*; AACC International: St. Paul, MN, USA, 2013.
37. European Brewery Convention. *Analytica. Method 4.16.1*, 5th ed.; Fachverlag Hans Carl: Nürnberg, Germany, 1998.
38. Martini, D.; Taddei, F.; Nicoletti, I.; Ciccoritti, R.; Corradini, D.; D'Egidio, M.G. Effects of genotype and environment on phenolic acids content and total antioxidant capacity in durum wheat. *Cereal Chem.* **2014**, *91*, 310–317. [CrossRef]
39. Nascimento, T.A.; Calado, V.; Carvalho, C.W.P. Effect of Brewer's spent grain and temperature on physical properties of expanded extrudates from rice. *LWT Food Sci. Technol.* **2017**, *79*, 145–151. [CrossRef]
40. Hidalgo, A.; Brandolini, A. Evaluation of heat damage, sugars, amylases and colour in beads from einkorn, durum and bread wheat flours. *J. Cereal Sci.* **2011**, *54*, 90–97. [CrossRef]
41. Erlandsson, A. Tritordeum. Evaluation of a New Food Cereal. Master's Thesis, Swedish University of Agricultural Sciences, Uppsala, Sweden, 2010.
42. Official Journal of the European Union. *REGULATION (EC) No 1924/2006 of the European Parliament and of the Council of 20 December 2006 on Nutrition and Health Claims Made on Foods*; Official Journal of the European Union: Brussels, Belgium, 2006.
43. European Food Safety Authority. Scientific opinion on the substantiation of a health claim related to oat beta glucan and lowering blood cholesterol and reduced risk of (coronary) heart disease pursuant to Article 14 of Regulation (EC) No 1924/200612010. *Efsa J.* **2010**, *8*, 1885.
44. Manthey, F.; Schorno, A. Physical and cooking quality of spaghetti made from whole wheat durum. *Cereal Chem.* **2002**, *79*, 504–510. [CrossRef]
45. Bustos, M.C.; Pérez, G.T.; León, A.E. Effect of four types of dietary fiber on the technological quality of pasta. *Food Sci. Technol. Int.* **2011**, *17*, 213–221. [CrossRef]
46. Foschia, M.; Peressini, D.; Sensidoni, A.; Brennan, C.S. The effects of dietary fibre addition on the quality of common cereal products. *J. Cereal Sci.* **2013**, *58*, 216–227. [CrossRef]
47. Foschia, M.; Peressini, D.; Sensidoni, A.; Brennan, M.A.; Brennan, C.S. How combinations of dietary fibres can affect physico-chemical characteristics of pasta. *LWT Food Sci. Technol.* **2015**, *61*, 41–46. [CrossRef]
48. Marconi, E.; Graziano, M.; Cubadda, R. Composition and utilization of barley pearling by-products for making functional pastas rich in dietary fiber and β-glucans. *Cereal Chem.* **2000**, *77*, 133–139. [CrossRef]
49. Cleary, L.; Brennan, C. The influence of a (1→3)(1→4)-β-d-glucan rich fraction from barley on the physico-chemical properties and in vitro reducing sugars release of durum wheat pasta. *Int. J. Food Sci. Technol.* **2006**, *41*, 910–918. [CrossRef]
50. Aravind, N.; Sissons, M.; Egan, N.; Fellows, C.M.; Blazek, J.; Gilbert, E.P. Effect of β-glucan on technological, sensory, and structural properties of durum wheat pasta. *Cereal Chem.* **2012**, *89*, 84–93. [CrossRef]

foods

MDPI

Article

Nutritional and Technological Quality of High Protein Pasta

Maria Cristina Messia, Francesca Cuomo, Luisa Falasca, Maria Carmela Trivisonno, Elisa De Arcangelis and Emanuele Marconi *

Department of Agricultural, Environmental and Food Sciences (DiAAA), University of Molise, Via F. De Sanctis snc, 86100 Campobasso, Italy; messia@unimol.it (M.C.M.); francesca.cuomo@unimol.it (F.C.); falasca@unimol.it (L.F.); mariacarmela.trivisonno@unimol.it (M.C.T.); elisa.dearcangelis@unimol.it (E.D.A.)

* Correspondence: marconi@unimol.it

Abstract: Pasta has an important role in human nutrition for its high content of complex carbohydrates and its widespread use. It can be an efficient delivery system or carrier of non-traditional raw material, including additional health-promoting ingredients. The partial replacement of semolina with high-protein raw materials leads to the improvement of the biological value of pasta proteins. In order to obtain pasta with high nutritional protein value and with excellent cooking properties, various recipes have been formulated with different percentages of semolina and unconventional high-protein raw materials (peas and soy isolate proteins, egg white, whey proteins and *Spirulina platensis*). High-protein pasta was produced using a pasta making pilot plant and the nutritional quality (protein content and quality) and sensorial properties were assessed. All experimental pastas showed optimal performances. Pasta prepared with pea protein isolate, whey proteins and *Spirulina platensis* showed improved chemical score and digestible indispensable amino acid scores, an eye-catching color, and an excellent cooking quality.

Keywords: pasta; unconventional ingredients; nutritional value; cooking quality; legumes; *Spirulina Platensis*

Citation: Messia, M.C.; Cuomo, F.; Falasca, L.; Trivisonno, M.C.; De Arcangelis, E.; Marconi, E. Nutritional and Technological Quality of High Protein Pasta. *Foods* **2021**, *10*, 589. https://doi.org/10.3390/foods10030589

Academic Editors: Francesca Nocente and Laura Gazza

Received: 12 February 2021
Accepted: 8 March 2021
Published: 11 March 2021

1. Introduction

Pasta is a foodstuff with an important role in human nutrition. It is popular with consumers for its easy handling, storage and preparation. In addition to its high content of complex carbohydrates with low glycaemic index, pasta also contains proteins. The quality of a protein is substantially related to its composition in essential amino acids and its digestibility. High-quality proteins contain all the essential amino acids at levels equal to or higher than those of the reference amino acid pattern of FAO/WHO/UN [1]. On the contrary, nutritionally incomplete or low biological value proteins are those lacking or deficient in one or more essential amino acids. All proteins of animal origin, with the exception of collagen, are considered complete proteins, while vegetable proteins, with a few exceptions, have a relative deficiency in certain essential amino acids. The limiting amino acid is the essential amino acid present in a protein in the lowest quantity, thus arginine is the limiting amino acid for casein, methionine for fish and egg proteins, and lysine is the amino acid of which commonly vegetable proteins are more deficient, especially those of cereals and cereal-based foods, such as pasta. In this context, the chemical score (CS) is the parameter used to describe the quality of proteins in terms of the potential ability of the dietary protein to provide the appropriate amount of essential amino acids. Proteins with a CS close or equal to 100 are considered better nutritionally, and therefore able to adequately meet human needs.

The interest in protein supplements is particularly exhibited by athletes and by people who need to bring a greater amount of protein into their diet. The protein requirement of the adult can increase significantly with sporting activity, starting from a base value, indicated in the LARN (Livelli di Assunzione di Riferimento di Nutrienti ed energia) [2], of 0.8 g/kg (protein/kg of body weight), useful for the sedentary subject, to values higher than 1.5 g/kg, indicated to cover the athlete's needs for the growth, maintenance, and repair of muscle [3–8]. However, it should be noted that adopting a diet very rich in proteins can lead to a reduction in carbohydrate intake, since sportspeople usually adhere to strict and scrupulous daily energy intakes. This imbalance in the diet could be counterproductive and potentially harmful in terms of health. There is a widespread belief among body builders that a high-protein diet, further integrated with purified proteins, is the fundamental factor for the development of muscle mass. The protein requirement increases if the training is aimed at developing strength and therefore muscle trophism, or if the training load is particularly intense. The use of nutritional supplements is very widespread among athletes of different levels, although the scientific literature does not report any data on their functions and their effects, which are instead promoted to the public. However, regardless of sportsmen, consumers are currently changing their eating habits thanks to the greater awareness they are acquiring regarding the well-being that a diet rich in foods of plant origin and low in foods of animal origin brings, both to the body and to the environment [9]. As shown by Ranganathan and coworkers [10], it is more expensive to obtain animal resources than plant-based ones. This aspect represents one of the main reasons for orienting new food styles towards other sustainable and effective sources able to provide high-quality food production while coping with population growth.

The commercial segment of high-protein pasta, driven by particular nutritional needs but also by food trends, has been growing in recent years. The nutritional needs required by consumers (higher protein content) have been met by the food industry, as demonstrated by the numerous products on the market. Innovative pasta recipes, including the replacement of semolina with alternative ingredients, have already been proposed, and non-traditional raw materials, soybean, pea, bean, chickpea flours or isolates, but also milk products such as whey proteins, casein and powdered milk, could be used [11–13]. In particular, legumes represent an interesting source of nutrients (protein, minerals, fiber) [14], and represent a valid ingredient in the development of diets that are healthy to humans and sustainable for the environment, since they can help to mitigate environmental climate change by reducing the carbon and water footprint [15]. Therefore, there has been an increasing interest in integrating legumes into food production. Although legumes' proteins are relatively low in sulfur amino acids and tryptophan, they have high lysine contents. Consequently, legumes and cereals are nutritionally complementary. The partial replacement of semolina with legume flours in the preparation of pasta leads to the improvement of the protein biological value and the amino acid CS.

More recently, microalgae and cyanobacteria such as *Chlorella* spp., *Dunaliella* spp., and *Spirulina* spp. are becoming more popular as new, highly nutritious food ingredients [16–18]. *Spirulina platensis* (spirulina) is rich in digestible protein, fat with unsaturated fatty acids, mineral, clorophyll and B group vitamins, in particular B12 vitamin [19,20]. For its characteristics, spirulina has been proposed for different food preparations, such as yogurt [21], snacks [22], or in Indian recipes [23] for the development of functional foods. Moreover, some studies have demonstrated the potential of this microalgae in the prevention and treatment of diseases related to metabolic syndrome [24]. Recently, a study has been published wherein spirulina has been used as a filling, together with other ingredients, in stuffed pasta [25]. The investigation was mainly focused on the acceptance of consumers of spirulina, and it emerged that its taste is accepted only in small amounts. Another recent study [26] used spirulina encapsulated in alginate microcapsules in the production of fresh pasta prepared with wheat flour. Encapsulation partially protects spirulina from the loss of its antioxidant potential, and the pasta presented green dots of a non-uniform color on the surface, which did not negatively influence the consumer's judgement.

Using legumes and spirulina as food ingredients represents an opportunity to reconcile the food system with the needs of the planet, and to encourage a healthy and balanced diet with beneficial effects for both humans and the environment, as indicated in the latest

European strategies on the agri-food system "From Farm to Fork Strategy—For a fair, healthy and environmentally friendly food system and Green Deal" [27].

However, the amount of high-protein material that can be added to or substituted for semolina represents a compromise between nutritional improvement and the achievement of satisfactory sensory and functional properties in pasta. Often, improving protein quantity and quality in pasta by the addition of various raw materials from plant or animal sources can lead to a decrease in pasta's sensory and cooking qualities.

Based on the above considerations, here, soy protein isolate, pea protein isolate, whey proteins, and spirulina were proposed as additional ingredients to improve the nutritional quality of semolina pasta. The ingredients were used to produce high-protein pasta, and the effects on the nutritional and cooking quality properties were investigated and compared with 100% semolina pasta.

2. Materials and Methods

2.1. Ingredients

Soy protein isolate (ABS FOOD srl, Peraga di Vigonza (PD), Italy), peas protein isolate (ABS FOOD srl, Peraga di Vigonza (PD), Italy), egg white (EUROVO srl, Bologna, Italy), whey proteins (Volac International Ltd., Hertfordshire, UK), high-quality durum wheat semolina (high protein, gluten index = 95) from a local distributor, and *Spirulina platensis* (spirulina) (ATI Biotech, Napoli, Italy) were used as ingredients to develop high-protein pasta.

2.2. Commercial Pasta

Five short commercial pastas ("rigatoni" and "penne" shape) with high protein contents were purchased in a local supermarket. The protein content and the ingredients listed on their label are shown in Table 1.

Table 1. Protein content and ingredients of commercial high-protein pasta (CP).

Sample	Protein (%)	Ingredients
CP1	60	soy protein isolate, wheat flour, gluten, egg white, pea protein, wheat fiber, inulin, guar gum
CP2	40	semolina, pea protein, egg
CP3	60	gluten, wheat flour, soy protein isolate, egg white powder, whey proteins, pea protein, wheat fiber, guar gum
CP4	52	vegetal proteins (soy, pea), semolina, egg white, sodium alginate, L-methionine, L-threonine
CP5	65.1	soy protein, lentil flour, pea protein, calcium caseinate, egg

2.3. Proximate Composition

Moisture, ash and lipid content were determined according to the ICC methods 109/1, 104/1 and 136, respectively [28]. The dietary fiber was determined according to the AACC method 32.05 [29]. Protein content (N $\times$ 6.25) was determined according to the Dumas combustion method (AACC method 46–30) [29], using a Leco nitrogen determiner, model FP 528 (Leco Corp., St. Joseph, MI, USA).

2.4. Amino Acids Analysis and Chemical Score

Amino acids were analyzed after acidic and alkaline hydrolysis. Acidic hydrolysis: a sample, corresponding to 25 mg of protein, was hydrolyzed with 25 mL of 6 N HCl at 110 °C for 24 h. Afterwards, the sample was cooled, filtered, evaporated to dryness and re-dissolved in 0.1 N HCl. Alkaline hydrolysis for tryptophan: a sample containing 10 mg of protein was added to 1 mL of distilled water, shaken, supplemented with 10 N NaOH (5 mL) and distilled water (4 mL), and then hydrolyzed for 18 h at 110 °C. After cooling, the sample was neutralized by adding 6 N HCl, evaporated to dryness, and re-dissolved in 0.1 N HCl. Before analysis, all samples were diluted 1:50–1:100 with ultra-pure water, and analyzed by an

ICS6000 chromatographic system (Thermo Fisher Scientific S.p.A, Milano, Italy). Separation was performed with an Aminopac PA10 analytical column (250 × 2 mm, 8.5 μm particle size) (Thermo Fisher Scientific S.p.A, Milano, Italy). Chromatographic separation of the amino acids was performed according to the following conditions (Table 2).

Table 2. Conditions for chromatographic separation of amino acids.

Mobile Phase (0.250 mL/min)				Time/Potential Waveform		
Time (min)	**H_2O (%)**	**NaOH (%)**	**NaOAc (%)**	**Time (sec)**	**Potential (V)**	**Integration**
0.0	80	20	0	0.00	+0.13	
2.0	80	20	0	0.04	+0.13	
12.0	80	20	0	0.05	+0.28	
16.0	68	32	0	0.11	+0.28	began
24.0	36	24	40	0.12	+0.60	
40.0	36	24	40	0.41	+0.60	
40.1	20	80	0	0.42	+0.28	
42.1	20	80	0	0.56	+0.28	end
42.2	80	20	0	0.57	−1.67	
62.0	80	20	0	0.58	−1.67	
				0.59	+0.93	
				0.60	+0.13	

Chemical score (CS) and digestible indispensable amino acid score (DIAAS%) were calculated according to the Food and Agriculture Organization [1] using the recommended amino acid scoring pattern for older children, adolescents and adults.

2.5. Pasta Making

Short pasta (rigatoncini shape) was manufactured through an experimental pasta making apparatus (NAMAD, Roma, Italy) composed of a press and a dryer following the approved method 66–41 [29]. The press (capacity 10–20 kg) was equipped with a vacuum-mixing and -extruding system, as well as with a water-cooling jacket for the barrel and the extrusion head to reduce heat and to maintain a constant extrusion temperature lower than 50 °C. The static dryer was equipped with a heat ventilator unit, to ensure uniform temperature and ventilation in all parts of the apparatus, and a moisture control unit. Semolina and other ingredients were mixed for 15 min with tap water (30 °C) to obtain a dough suitable for extrusion. Extrusion occurred at 30 ± 2 °C and at a pressure of 76 ± 5 bar. Each series of short pasta was dried at a maximum temperature of 50 °C for 24 h. At the end of the drying cycle pasta was conditioned at room temperature (~20 °C) for 24 h.

2.6. Pasta Characterization

Optimum cooking time, firmness (by chewing), liveliness (by manual handling) and starch release (by manual handling) were determined according to International Standard ISO 7304-1 [30]. A rating scale ranging from 10 to 100 was used (Table 3). A panel of 10 trained judges was used to assess pasta characteristics. The total score was calculated by adding together the ratings obtained for firmness, liveliness and starch release and then dividing the sum by 3.

2.7. Statistical Analysis

Data reported for all parameters are the average values of measurements obtained from the analysis of three different aliquots of each sample, and were expressed as mean ± standard deviation (mean value ± sd). Analysis of variance (ANOVA) and Tukey HSD tests were performed on the protein content and cooking quality scores of the experimental pastas using RStudio version 1.2.5033 (RStudio Team (2019). RStudio: Integrated Development for R. RStudio, Inc., Boston, MA, USA, http://www.rstudio.com/ (accessed on 3 March 2021)). Significant differences were set for $p < 0.05$.

Table 3. Rating scale for pasta sensory analysis [30].

Firmness	Liveliness	Starch Release
100—very high (very firm)	100—very high (not at all sticky)	100—very low (no starch)
80—high	80—high	80—low
50—medium	50—medium	50—medium
30—low	30—low	30—high
10—very low (very tender)	10—very low (very sticky)	10—very high (large quantity of starch)

3. Results

3.1. Cooking Quality Assessment of Commercial High-Protein Pasta

The nutritional needs of consumers are met by the food industry, as demonstrated by numerous products on the market. Five commercial pastas with a high protein contents (ranging between 40 and 65%) were checked for composition and ingredients. As shown in the list of ingredients (Table 1), the abovementioned pastas included a wide range of ingredients such as soy protein, pea protein, lentils flour, eggs, egg white, inulin, caseinate and gluten, in addition to the presence of additives such as E412 (guar gum) and E401 (sodium alginate), and it emerged that they were prepared by reducing or annihilating the percentage of durum wheat semolina used in the formulation. Moreover, food additives, such as guar gum are commonly used as structuring agents with the aim of replacing the gluten that is missing in the alternative ingredients used to increase the protein content. Although commercial pastas satisfy the requirement related to the high protein content, they are not always able to satisfy the consumer from a sensorial point of view.

3.2. Characterization of Raw Materials for High-Protein Pasta Production

Durum wheat semolina, due to the rheological properties of its proteins (gluten) and the high content of pigments, is considered the best raw material for pasta making. The preparation of pasta with unconventional ingredients is challenging due to the absence/reduced formation of the protein network that prevents the disintegration of the pasta during cooking. In this experimentation, the possibility of using semolina in combination with other high protein raw materials to produce pasta with high protein content, improved protein quality and optimum cooking quality was studied. In order to only partially replace the semolina, among the possible ingredients, we tested protein isolates of soy and pea, egg white, whey proteins and spirulina.

To obtain excellent quality pasta, without the adjuvants/additives (e.g., guar gums, sodium alginate) generally enclosed in the recipes of commercial pasta, high-protein and high-gluten semolina was used. To counteract the lack of gluten and help the formation of a cohesive mass, we chose to add egg white and whey proteins, as widely reported in the literature [31,32].

The proximate composition of semolina and unconventional raw materials used for the production of experimental high-protein pasta is reported in Table 4.

For the selected raw material, the amino acids content, the CS, and the definition of the limiting amino acid, calculated on the basis of the Food and Agriculture Organization's described pattern [1], were assessed (Table 5).

Table 4. Proximate composition (g/100 g d.w.) of wheat semolina and other raw materials.

Sample	Protein	Lipid	Ash	Carbohydrates *	Fiber
Wheat Semolina	14.3 ± 0.01	1.5 ± 0.02	0.8 ± 0.00	79.2	4.2 ± 0.30
Soy Protein Isolate	90.0 ± 0.21	1.1 ± 0.01	6.5 ± 0.11	1.3	1.1 ± 0.23
Pea Protein Isolate	87.9 ± 0.05	2.0 ± 0.02	6.9 ± 0.21	1.4	1.8 ± 0.43
Egg White	89.2 ± 0.12	0.0 ± 0.00	8.3 ± 0.05	0.0	0.0 ± 0.00
Whey Proteins	91.8 ± 0.81	0.4 ± 0.00	2.1 ± 0.23	5.7	0.0 ± 0.00
Spirulina	58.5 ± 0.32	8.9 ± 0.15	7.3 ± 0.45	20.8	4.5 ± 0.30

* Calculated by difference.

Table 5. Amino acid (g/100 g protein), chemical score (CS) and limiting amino-acid of soy protein, peas protein, egg white, whey proteins and wheat semolina.

Essential Amino Acids	Soy Protein Isolate	Peas Protein Isolate	Egg White	Whey Proteins	Wheat Semolina	FAO [1] Amino Acid Scoring Patterns (mg/g)
Histidine	2.29 ± 0.13	3.50 ± 0.09	2.20 ± 0.05	2.13 ± 0.03	2.05 ± 0.11	16
Isoleucine	3.92 ± 0.22	1.65 ± 0.08	5.35 ± 0.09	7.12 ± 0.10	3.69 ± 0.07	30
Leucine	7.49 ± 0.01	7.80 ± 0.08	8.06 ± 0.30	11.35 ± 0.55	7.07 ± 0.05	61
Lysine	5.69 ± 0.14	6.63 ± 0.31	6.78 ± 0.21	9.83 ± 0.13	2.09 ± 0.24	48
Methionine	1.20 ± 0.04	0.25 ± 0.22	3.94 ± 0.10	2.25 ± 0.09	1.56 ± 0.08	Methionine + Cysteine 23
Cysteine	1.84 ± 0.04	1.48 ± 0.09	2.92 ± 0.05	2.51 ± 0.21	2.45 ± 0.07	
Phenylalanine	4.86 ± 0.15	5.99 ± 0.12	5.85 ± 0.15	3.44 ± 0.28	4.89 ± 0.19	Phenylalanine + Tyrosine 41
Tyrosine	3.41 ± 0.22	2.68 ± 0.08	4.19 ± 0.19	3.42 ± 0.56	2.56 ± 0.09	
Threonine	3.77 ± 0.00	3.19 ± 0.10	4.67 ± 0.09	7.72 ± 0.02	2.77 ± 0.04	25
Valine	3.98 ± 0.07	4.64 ± 0.13	6.84 ± 0.10	6.65 ± 0.23	4.33 ± 0.01	40
Tryptophan	0.51 ± 0.02	1.15 ± 0.09	1.60 ± 0.07	1.93 ± 0.04	0.98 ± 0.02	6.6
Chemical Score (CS)	77	79	100	100	44	
Limiting Amino acid	Tryptophan	Sulphur amino acid (Met + Cys)	–	–	Lysine	

3.3. Definition of Pasta Recipes and Pasta Cooking Quality

Different formulations for high-protein pasta were hypothesized in order to produce pasta with a protein content above 40% and an improved CS (90–100%) compared to 100% semolina pasta. The formulations were studied, taking into account the costs of raw materials, and including ingredients at appropriate quantities to obtain a final product with acceptable sensorial characteristics and good cooking quality. Moreover, salt was not added to ensure that the salt in the pasta was due only to the sodium naturally contained in the ingredients, and also to meet the needs of consumers as regards preventing arterial hypertension. Spirulina (1%) as a source of proteins and color was also included among the ingredients for pasta production (formulation FHP2, FHP4, FHP6 and FHP8).

The hypothesized formulations are shown in Table 6.

High-protein short pastas ("rigatoncini" shape) were produced, starting from the formulations in Table 6. The 100% semolina pasta was used as the control sample. The results for the protein content, chemical score, DIAAS%, optimal cooking time and cooking quality of all produced pastas are reported in Table 7.

Table 6. Hypothesized high-protein pasta formulations (FHP).

Formulation	Ingredients (%)					
	Wheat Semolina	Soy Protein Isolate	Pea Protein Isolate	Egg White	Whey Proteins	Spirulina
FHP0	100	0	0	0	0	0
FHP1	44	17.4	30.8	7.8	0	0
FHP2	43	17.4	30.8	7.8	0	1
FHP3	38	25	25	12	0	0
FHP4	37	25	25	12	0	1
FHP5	45	23	20	12	0	0
FHP6	44	23	20	12	0	1
FHP7	60	0	30	0	10	0
FHP8	59	0	30	0	10	1

Table 7. Cooking quality of experimental high-protein pastas (Pasta HP).

Pasta	Protein (%)	Chemical Score	DIAAS (%)	Optimal Cooking Time (min)	Cooking Quality				Panel Comments
					Firmness	Liveliness	Starch Release	Total Score	
Pasta HP0	12.4 ± 0.01 [h]	44	36	13′:00″	100 ± 0.0 [a]	90 ± 4.71 [a]	90 ± 4.08 [b]	93.3	Amber/yellow color, pleasant smell and taste, optimal consistency
Pasta HP1	50.6 ± 0.02 [c]	91	100	14′:00″	100 ± 0.0 [a]	80 ± 5.27 [b]	80 ± 3.33 [c]	86.7	Light brown color, pea flavor, excellent consistency
Pasta HP2	50.9 ± 0.04 [b]	91	100	14′:50″	100 ± 0.0 [a]	90 ± 4.08 [a]	80 ± 3.33 [c]	90.0	Green color, pea smell, excellent consistency
Pasta HP3	54.5 ± 0.10 [a]	100	100	15′:00″	80 ± 2.36 [c]	80 ± 5.27 [bc]	90 ± 6.24 [b]	83.3	Light brown color, mild pea flavor, good consistency
Pasta HP4	54.7 ± 0.05 [a]	100	100	15′:00″	90 ± 7.07 [b]	85 ± 4.71 [ac]	90 ± 4.08 [b]	88.3	Green color, slight herbaceous flavor, good consistency
Pasta HP5	50.1 ± 0.11 [e]	95	100	14′:30″	100 ± 0.0 [a]	85 ± 4.71 [ac]	80 ± 4.71 [c]	88.3	Light brown color, pleasant taste, optimal consistency
Pasta HP6	50.4 ± 0.05 [d]	95	100	15′:00″	100 ± 0.0 [a]	87 ± 5.87 [a]	80 ± 2.36 [c]	89.0	Green color, pleasant flavor, optimal consistency
Pasta HP7	40.3 ± 0.00 [g]	100	100	15′:30″	97.5 ± 4.24 [a]	90 ± 4.08 [a]	100 ± 0.00 [a]	95.8	Light brown color, optimal consistency
Pasta HP8	40.7 ± 0.01 [f]	100	100	15′:30″	98 ± 2.58 [a]	92 ± 5.87 [a]	100 ± 0.00 [a]	96.7	Green color, optimal consistency

Different letters in a column indicate statistically significant differences ($p < 0.05$).

4. Discussion

The nutritional quality of a protein that is deficient in essential amino acids can be improved by suitable supplementation with other proteins rich in essential amino acids. Therefore, the addition of proteins from other sources in cereal-based formulations results in a complete and balanced level of essential amino acids.

The heterogeneity of raw materials (protein can come from both plant and animal sources) potentially usable in the production of cereal-based foods, and the replacement of all or part of the conventional flours with other cereals or ingredients different from cereals, often entails the need to make changes to the traditional production process. Balanced formulations and adequate technological processes must be adopted to compensate for any changes in functional properties caused by the incorporation of new ingredients [12,32,33].

The introduction of unconventional material in the recipe of commercial pasta (Table 1) definitely increased the protein content, especially if these are added at high levels. However, this impacted the quality attributes of pasta. In fact, besides a high protein content, the panelists' observations about smell, taste, texture and color were quite negative. To overcome this problem, with the aim of achieving optimal technological behavior that could have a positive influence on the cooking quality of the pasta, the formulations for innovative pasta were hypothesized, maintaining a higher percentage of semolina than the remaining ingredients (Table 6). The loss of firmness following the cooking caused by gluten deficiency, and the possible presence of unpleasant tastes and flavors due to alternative ingredients such as legumes, would prejudice the consumer's acceptance of the pasta. In fact, pasta compounds such as proteins, fat, and carbohydrates can absorb legumes' flavor compounds, resulting in their retention [34]. It thus means that the amount of pulse ingredient to add will be limited by flavor characteristics. Up to a certain percentage, the "off" flavor of the pulse ingredient can be masked by other compounds present in the food matrix [35].

To assess the improvement of the nutritional characteristics and the effects of the unconventional raw material's addition on the cooking quality of pasta, the protein content, the CS and the cooking quality were evaluated. Data in Table 7 show that all the experimental pasta had good protein contents (40.7–54.7% fresh weight, f.w.), an improved CS (CS = 91–100) compared to the pasta with 100% semolina (CS = 44), and excellent cooking quality (total score between 83.3 and 96.7). To assess the protein quality of pasta, besides CS, the DIAAS% was calculated. The actual capacity of protein to satisfy the amino acid needs requires the use of corrections for amino acid digestibility and availability. The FAO [1] recommendation is to use DIAAS as the measure of protein quality, rather than measures such as the protein efficiency ratio (PER). A nutritional claim for protein content (i.e., "source", "high" according to Regulation (EC) No 1924/2006 [36]) should be coupled to the computing of the DIAAS values, to discriminate the quality of the protein itself. For excellent/high-protein quality, DIAAS $\geq$100 were proposed, for good/source values ranging from 75 to 99, while it was stated that no claim should be allowed for the cut-off value of, e.g., 75. All the recipes proposed in this experimentation led to the production of pasta with DIAAS higher than 100, showing the good combinations used to achieve high-quality protein in the final product.

The presence of a high content of semolina, egg white and whey proteins ensured the structuring of the pasta and therefore the cooking quality, achieving an evaluation comparable to the 100% semolina pasta used as a control. Pasta produced using formulation FPH7 and 8 in addition to having a protein content of 41.1% f.w. and optimal protein quality (CS = 100, DIAAS = 100) higher than the control, among all the experimental pasta, showed the highest score for cooking quality. The results have shown that it is possible to produce nutritionally valid pasta with excellent cooking quality by only partially replacing the semolina and without adding adjuvants/additives, as used in and shown on most commercial pasta labels. Moreover, according to Regulation (EC) No 1924/2006 [36] and Commission Regulation (EU) No 432/2012 [37], all the pastas of this experimentation can boast both the nutrition claims (source of protein, high protein) and the health claims related to proteins, because they are foods that are at least a source of protein, as referred to in the claim source of protein, as listed in the Annex to Regulation (EC) No 1924/2006 [36].

Moreover, the presence of spirulina not only contributed to increasing the protein content, but in combination with pea protein and whey proteins, also determined a higher pasta firmness. According to Fradique et al. [38], the reinforcement of the gluten network could cause an extra establishment of disulfide bonds, formed between the sulfhydryl groups of cysteine residues in gluten proteins. Moreover, all pasta produced with spirulina showed an intense green color (Figure 1), which is less usual for pasta consumers, but this did not affect its acceptability because its taste and flavor were ordinary. The intense green color maintained by the pasta after the cooking (Figure 1) is also due to the fact that spirulina presents only chlorophyll *a* in its constitution, which is more stable under thermal processes than the chlorophyll *b* of vegetables, including the spinach generally used to color the products [39]. Similar observations have also recently been reported by Mostolizadeh et al. [40] using spirulina in pasta production.

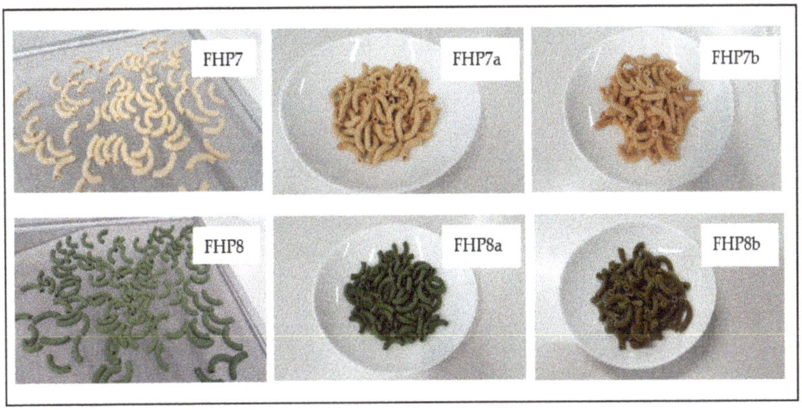

Figure 1. Raw (a) and cooked pasta (b) produced using formulation 7 (FHP7) and 8 (FHP8).

5. Conclusions

The study highlights the suitability of unconventional raw materials, such as legumes, whey proteins and spirulina, for obtaining pastas improved in terms of protein content, amino acids chemical score and cooking quality.

Good quality semolina (high protein, gluten strength) and its limited substitution with other ingredients is the key to obtaining pasta with a high cooking quality. Through suitable formulations it is possible to obtain not only the protein-enriched pasta, but also the amino acid complementarity useful for carrying out physiological functions and for ensuring the production of pasta with high-quality proteins. Spirulina, although in small concentrations, contributed to improving protein content, positively affecting the pasta firmness and leading to a green color in the product, which was stable even after processing and cooking.

Examples of fortified pasta produced in this experimentation may help to broaden the offer for people who want to improve the nutritional quality of their diet or satisfy the particular needs of sportsmen.

Author Contributions: Conceptualization, M.C.M. and E.M.; formal analysis, L.F., M.C.T., E.D.A.; data curation, M.C.M., F.C.; writing—original draft preparation, M.C.M., F.C., E.D.A.; writing—review and editing, M.C.M., E.M., F.C.; supervision, M.C.M., E.M.; funding acquisition, M.C.M., E.M. All authors have read and agreed to the published version of the manuscript.

Funding: This research received no external funding.

Informed Consent Statement: Not applicable.

Data Availability Statement: The data presented in this study are available on request from the corresponding author.

Conflicts of Interest: The authors declare no conflict of interest.

References

1. FAO/WHO. Dietary protein quality evaluation in human nutrition. *FAO Food Nutr. Pap.* **2013**, *92*, 1–66.
2. Società Italiana di Nutrizione Umana. Livelli di Assunzione di Riferimento di Nutrienti ed Energia per la Popolazione Italiana IV Revisione (LARN 2014). Available online: https://sinu.it/tabelle-larn-2014/ (accessed on 12 January 2021).
3. Coleman, E. Protein requirements for athletes. *Clin. Nutr. Insight* **2012**, *38*, 1–3. [CrossRef]

4. Lemon, P.W.; Tarnopolsky, M.A.; MacDougall, J.D.; Atkinson, S.A. Protein requirements and muscle mass/strength changes during intensive training in novice bodybuilders. *J. Appl. Physiol.* **1992**, *73*, 767–775. [CrossRef]
5. Moore, D.R.; Robinson, M.J.; Fry, J.L.; Tang, J.E.; Glover, E.I.; Wilkinson, S.B.; Prior, T.; Tarnopolsky, M.A.; Phillips, S.M. Ingested protein dose response of muscle and albumin protein synthesis after resistance exercise in young men. *Am. J. Clin. Nutr.* **2009**, *89*, 161–168. [CrossRef]
6. Poortmans, J.R.; Carpentier, A.; Pereira-Lancha, L.O.; Lancha, A., Jr. Protein turnover, amino acid requirements and recommendations for athletes and active populations. *Braz. J. Med. Biol. Res.* **2012**, *45*, 875–890. [CrossRef]
7. Tang, J.E.; Phillips, S.M. Maximizing muscle protein anabolism: The role of protein quality. *Curr. Opin. Clin. Nutr. Metab. Care* **2009**, *12*, 66–71. [CrossRef]
8. Tarnopolsky, M.A.; MacDougall, J.D.; Atkinson, S.A. Influence of protein intake and training status on nitrogen balance and lean body mass. *J. Appl. Physiol.* **1988**, *64*, 187–193. [CrossRef] [PubMed]
9. Pfeiffer, C.; Speck, M.; Strassner, C. What leads to lunch—How social practices impact (non-) sustainable food consumption/eating habits. *Sustainability* **2017**, *9*, 1437. [CrossRef]
10. Ranganathan, J.; Vennard, D.; Waite, R.; Dumas, P.; Lipinski, B. Shifting Diets for a Sustainable Food Future. Available online: http://www.indiaenvironmentportal.org.in/files/file/Shifting_Diets_for_a_Sustainable_Food_Future.pdf (accessed on 6 December 2020).
11. Foschia, M.; Horstmann, S.W.; Arendt, E.K.; Zannini, E. Legumes as functional ingredients in gluten-free bakery and pasta products. *Annu. Rev. Food Sci. Technol.* **2017**, *8*, 75–96. [CrossRef] [PubMed]
12. Marconi, E.; Messia, M.C. Pasta made from non traditional raw materials: Technological and nutritional aspects. In *Durum Wheat Chemistry and Technology*; AACC International Press: St Paul, MN, USA, 2012.
13. Verardo, V.; Gómez-Caravaca, A.M.; Messia, M.C.; Marconi, E.; Caboni, M.F. Development of functional spaghetti enriched in bioactive compounds using barley coarse fraction obtained by air classification. *J. Agric. Food Chem.* **2011**, *59*, 9127–9134. [CrossRef] [PubMed]
14. Asif, M.; Rooney, L.W.; Ali, R.; Riaz, M.N. Application and opportunities of pulses in food system: A review. *Crit. Rev. Food Sci. Nutr.* **2013**, *53*, 1168–1179. [CrossRef] [PubMed]
15. Boukid, F.; Zannini, E.; Carini, E.; Vittadini, E. Pulses for bread fortification: A necessity or a choice? *Trends Food Sci. Technol.* **2019**, *88*, 416–428. [CrossRef]
16. Kadam, S.U.; Prabhasankar, P. Marine foods as functional ingredients in bakery and pasta products. *Food Res. Int.* **2010**, *43*, 1975–1980. [CrossRef]
17. Matos, Â.P. The impact of microalgae in food science and technology. *J. Am. Oil Chem. Soc.* **2017**, *94*, 1333–1350. [CrossRef]
18. Vendruscolo, R.G.; Fagundes, M.B.; Jacob-Lopes, E.; Wagner, R. Analytical strategies for using gas chromatography to control and optimize microalgae bioprocessing. *Curr. Opin. Food Sci.* **2019**, *25*, 73–81. [CrossRef]
19. Ciferri, O.; Tiboni, O. The biochemistry and industrial potential of Spirulina. *Annu. Rev. Microbiol.* **1985**, *39*, 503–526. [CrossRef]
20. Shimamatsu, H. Mass production of Spirulina, an edible microalga. *Hydrobiologia* **2004**, *512*, 39–44. [CrossRef]
21. da Silva, S.C.; Fernandes, I.P.; Barros, L.; Fernandes, Â.; Alves, M.J.; Calhelha, R.C.; Pereira, C.; Barreira, J.C.M.; Manrique, Y.; Colla, E. Spray-dried Spirulina platensis as an effective ingredient to improve yogurt formulations: Testing different encapsulating solutions. *J. Funct. Foods* **2019**, *60*, 103427. [CrossRef]
22. Chacón-Lee, T.L.; González-Mariño, G.E. Microalgae for "healthy" foods—possibilities and challenges. *Compr. Rev. Food Sci. Food Saf.* **2010**, *9*, 655–675. [CrossRef]
23. Iyer, U.M.; Dhruv, S.A.; Mani, I.U. Spirulina and its therapeutic implications as a food product. In *Spirulina Human Nutrition and Health*; Belay, A., Ed.; Crc Press, Taylor & Francis Publishing group: Boca Raton, FL, USA, 2007; pp. 51–70.
24. Lafarga, T.; Fernández-Sevilla, J.M.; González-López, C.; Acién-Fernández, F.G. Spirulina for the food and functional food industries. *Food Res. Int.* **2020**, *137*, 109356. [CrossRef] [PubMed]
25. Grahl, S.; Strack, M.; Mensching, A.; Mörlein, D. Alternative protein sources in Western diets: Food product development and consumer acceptance of spirulina-filled pasta. *Food Qual. Prefer.* **2020**, *84*, 103933. [CrossRef]
26. Zen, C.K.; Tiepo, C.B.V.; da Silva, R.V.; Reinehr, C.O.; Gutkoski, L.C.; Oro, T.; Colla, L.M. Development of functional pasta with microencapsulated Spirulina: Technological and sensorial effects. *J. Sci. Food Agric.* **2020**, *100*, 2018–2026. [CrossRef]
27. European Commission. Farm to Fork Strategy–For a Fair, Healthy and Environmentally-Friendly Food System. 2020. Available online: https://ec.europa.eu/food/sites/food/files/safety/docs/f2f_action-plan_2020_strategyinfo_en.pdf (accessed on 2 December 2020).
28. ICC. *Standard Methods of the International Association for Cereal Science and Technology*; ICC: Vienna, Austria, 1995.
29. AACC. *American Association of Cereal Chemists. Approved Methods of AACC*, 10th ed.; The Association: St. Paul, MN, USA, 2000.
30. ISO. International Standard ISO 7304-1:2016, Durum Wheat Semolina and Alimentary Pasta—Estimation of Cooking Quality of Alimentary Pasta by Sensory Analysis. Available online: https://www.iso.org/standard/44312.html (accessed on 1 January 2021).
31. Laleg, K.; Barron, C.; Santé-Lhoutellier, V.; Walrand, S.; Micard, V. Protein enriched pasta: Structure and digestibility of its protein network. *Food Funct.* **2016**, *7*, 1196–1207. [CrossRef]
32. Marti, A.; Pagani, M.A. What can play the role of gluten in gluten free pasta? *Trends Food Sci. Technol.* **2013**, *31*, 63–71. [CrossRef]

33. Petitot, M.; Boyer, L.; Minier, C.; Micard, V. Fortification of pasta with split pea and faba bean flours: Pasta processing and quality evaluation. *Food Res. Int.* **2010**, *43*, 634–641. [CrossRef]
34. Fischer, N.; Widder, S. How proteins influence food flavor: The chemistry of flavor interactions. *Food Technol.* **1997**, *51*, 68–70.
35. Roland, W.S.; Pouvreau, L.; Curran, J.; van de Velde, F.; de Kok, P.M. Flavor aspects of pulse ingredients. *Cereal Chem.* **2017**, *94*, 58–65. [CrossRef]
36. European Commission. Regulation (EC) No. 1924/2006 of the European Parliament and of the Council of 20 December 2006 on nutrition and health claims made on foods. *Off. J. Eur. Union* **2006**, *L404*, 9–25.
37. European Commission. Regulation (EC) No. 432/2012 of 16 May 2012 establishing a list of permitted health claims made on foods, other than those referring to the reduction of disease risk and to children's development and health. *Off. J. Eur. Union* **2012**, *L136*, 1–40.
38. Fradique, M.; Batista, A.P.; Nunes, M.C.; Gouveia, L.; Bandarra, N.M.; Raymundo, A. Incorporation of Chlorella vulgaris and Spirulina maxima biomass in pasta products. Part 1: Preparation and evaluation. *J. Sci. Food Agric.* **2010**, *90*, 1656–1664. [CrossRef] [PubMed]
39. Lemes, A.C.; Takeuchi, K.P.; de Carvalho, J.C.M.; Danesi, E.D.G. Fresh pasta production enriched with Spirulina platensis biomass. *Braz. Arch. Biol. Technol.* **2012**, *55*, 741–750. [CrossRef]
40. Mostolizadeh, S.S.; Moradi, Y.; Mortazavi, M.S.; Motallebi, A.A.; Ghaeni, M. Effects of incorporation Spirulina platensis (Gomont, 1892) powder in wheat flour on chemical, microbial and sensory properties of pasta. *Iran. J. Fish. Sci.* **2020**, *19*, 410–420.

foods

MDPI

Article

Potential Application of Resistant Starch Sorghum in Gluten-Free Pasta: Nutritional, Structural and Sensory Evaluations

Mariasole Cervini [1], Alice Gruppi [2], Andrea Bassani [2], Giorgia Spigno [2] and Gianluca Giuberti [2,*]

1 Department of Biotechnology, University of Verona, Strada Le Grazie 15, 37134 Verona, Italy; mariasole.cervini@univr.it

2 Department for Sustainable Food Process (DiSTAS), Università Cattolica del Sacro Cuore, via Emilia Parmense 84, 29122 Piacenza, Italy; alice.gruppi@unicatt.it (A.G.); andrea.bassani@unicatt.it (A.B.); giorgia.spigno@unicatt.it (G.S.)

* Correspondence: gianluca.giuberti@unicatt.it

Abstract: Gluten-free (GF) pasta samples containing rice flour replaced with 0, 5, 10, 15 g/100 g (*w*/*w*) of a resistant starch ingredient from annealed sorghum starch (annRS) were formulated. The highest total dietary fiber and RS contents ($p < 0.05$) were measured in uncooked pasta with 15 g/100 g of annRS addition (15-annRS). After cooking, the 15-annRS pasta was characterized by an RS content of 5.8 g/100 g dry matter, confirming the thermal resistance of annRS. The use of annRS positively influenced the optimal cooking time, the cooking loss, the firmness, and the stickiness of the cooked samples, with not remarkably change in color after cooking. The starch hydrolysis index values decreased as the level of annRS increased. Despite a significant decrease in the overall sensory with increasing levels of annRS, all samples were characterized by a value > 5, which is considered the limit of acceptability. The use of annRS in GF pasta up to 15 g/100 g can contribute to creating GF products with high total dietary fiber content, slowly digestible starch properties, and without drastically compromising the sensory attributes.

Keywords: resistant starch; hydrolysis index; dietary fibre; pasta; annealing

Citation: Cervini, M.; Gruppi, A.; Bassani, A.; Spigno, G.; Giuberti, G. Potential Application of Resistant Starch Sorghum in Gluten-Free Pasta: Nutritional, Structural and Sensory Evaluations. *Foods* **2021**, *10*, 908. https://doi.org/10.3390/foods10050908

Academic Editors: Laura Gazza and Francesca Nocente

Received: 7 April 2021
Accepted: 20 April 2021
Published: 21 April 2021

1. Introduction

Cereal-based gluten-free (GF) products are not only exclusively consumed by individuals suffering from medically diagnosed coeliac disease but by a growing number of consumers who spontaneously reduce and/or avoid gluten from their eating habits [1]. However, divergences regarding the nutritional quality of cereal-based GF foods compared to gluten-containing counterparts are still present [2]. In particular, data comparison of the nutritional composition of cereal-based GF alternatives to gluten-containing foods generally indicates lower dietary fiber content, higher glycaemic index, and higher total fat content [3–5]. The nutritional imbalance of GF cereal-based products may also contribute to weight gain and related metabolic diseases for individuals following a strict GF diet [6,7]. Starting from these considerations, research has been conducted to ameliorate the nutritional profile of different cereal-based GF products, including dry pasta.

Overall, dry pasta is considered a suitable product to reformulate as GF, aiming to improve the nutritional profile [8]. In this context, one of the most flexible strategies is the partial replacement of common GF flours and/or starches with novel nutrition-dense ingredients [1,2,8,9]. In regards to raw materials, the potential use of ingredients rich in resistant starch (RS) is gaining importance in making GF pasta [5,10–12].

The RS fraction is that fraction of starch that escapes digestion in the small intestine to be fermented in the large intestine favoring a series of health-related benefits comparable to those of dietary fiber [13,14]. Besides, increasing the RS amount in pasta may result in lower glycaemic carbohydrate content and lower in vitro starch digestibility [15], even if discrepancies exist, probably related to the type and properties of the RS used in the formulation [10,11,15,16]. In this sense, as a function of the inherent heat stability of the RS, the pasta manufacturing process along with the cooking step can destroy most forms of RS [11,17,18]. This gives importance to the search for ingredients containing thermally stable forms of RS and to the need to evaluate their functionality and effect on pasta formulation.

Several attempts have been made to generate heat-stable RS ingredients through physical, chemical, and enzymatic treatments of different native starches [19,20]. Besides, there is a need to find alternative underutilized RS sources for possible food applications In this context, promising results have been reported by subjecting isolated white sorghum starch to annealing (annRS) [19]. The authors reported that the resulting annRS had a high RS content and greater heat stability to the native starch form. This novel RS-rich ingredient has been so far tested in GF biscuit formulation [21]. In particular, the use of annRS up to 45 g/100 g in the recipe contributed to the formulation of products with high RS content, slowly digestible starch properties, and without compromised quality and sensory attributes [21]. However, to the best of our knowledge, information concerning its functionality in GF dry pasta is lacking.

To better explore the potential of the annRS in GF dry pasta, GF pasta was formulated by replacing rice flour with increasing levels of annRS (up to 15 g/100 g) in the recipe. Newly developed products were evaluated for their RS content prior to and after cooking, along with the in vitro starch digestibility on cooked samples. Sensory analysis was also conducted to explore if the use of annRS could play a role in modifying the sensory attributes.

2. Materials and Methods

2.1. Raw Materials

White rice flour was supplied from Pedon S.p.A. (Molvena, Italy). As reported in the label, the chemical composition was moisture content 7.7 g/100 g; crude lipid 1.6 g/100 g; total starch 75.9 g/100 g; total sugar 0.5 g/100 g; crude protein 7.1 g/100 g, total dietary fiber 1.5 g/100 g of product. The particle size of the rice flour was <0.2 mm. The annRS ingredient was obtained from annealed white sorghum starch as previously detailed [19]. White sorghum starch was firstly isolated from commercial dehulled white sorghum flour ((*Sorghum bicolor* (L.) Moench)) purchased from CiboCrudo s.r.l. (Roma, Italy) and then dispersed in distilled water (ratio of 1:4 *w*/*v* starch to water) at 50 °C for 24 h under constant agitation. The liquid fraction was removed after centrifugation (4000 rpm; 15 min), and the remaining solid residue containing the annRS was oven-dried at 40 °C for 12 h (final moisture content of 8.7 g/100 g) and finely ground (0.5 mm screen; Retsch ZM1; Brinkman Instruments, Rexdale, ON, Canada). The RS content of the annRS was 53.5 g/100 g dry matter (DM), whereas the RS of white rice flour was 1.7 g/100 g DM.

2.2. Pasta Preparation

Macaroni-shaped GF pasta was produced in a customized plant installation (about 12 kg/h) consisting of a mixer, an extruder, and a cabinet dryer. The control GF pasta recipe contained rice flour (99.5 g/100 g dry flour basis) and mono- and di-glycerides of fatty acids (0.5 g/100 g; E471; Lucgel S.r.L, Perugia, Italy) (control). The annRS-enriched blends containing 5, 10, and 15 g/100 g *w*/*w* annRS were produced by replacing rice flour with the corresponding annRS level (5-annRS, 10-annRS, and 15-annRS, respectively). For each formulation, dry flour blends (6 kg) and tap 37 °C water were mixed (13 min; Procut Omni20, Inox-Fer s.r.l., Reggio Emilia, Italy) to obtain a uniform hydrated mass with a final water content of 35 g/100 g. The hydrated mass was heated in the mixer by steam at 0.3 MPa at 120 °C for 15 min to induce starch gelatinization. Then, it was formed in a single-screw extruder with a bronze macaroni-shaped die under vacuum conditions (La Parmigiana model RZ50, Parma, Italy) by keeping the dough temperature < 50 °C. The auger extrusion speed was 20 rpm. Samples were dried at 50 °C for 14 h in a cabinet dryer (La Parmigiana model ESS20, Parma, Italy). The control pasta was prepared under the same conditions. Dried GF pasta samples were stored at room temperature until analyzed. For each recipe, two batches were produced. The highest level of inclusion of annRS in the recipe was selected considering preliminary trials. Going beyond this level caused difficulties associated with the extrusion process.

2.3. Chemical Composition

Dry pasta was analyzed for proximate composition, including DM, ash, crude protein, crude lipid, and total starch [22]. The total dietary fiber (TDF) content was assessed enzymatically (Megazyme assay kit K-INTDF 02/15). This assay kit includes RS in the assessment of the TDF content in foods. A commercial assay kit (K-RSTAR 02/17, Megazyme International, Wicklow, Ireland) was used for the quantification of RS in both uncooked and cooked samples following manufacturer instructions. For cooked samples, 50 g of GF pasta were boiled in distilled water to optimal cooking time (OCT; see the specific paragraph), treated with liquid nitrogen, and lyophilized (method 2002.02) [22]. Samples were ground through a 0.5-mm screen. The apparent RS retention (aRSr) was calculated as follows:

$$\text{aRSr} = \text{RS in cooked sample (g/100 g dry weight)/RS in uncooked sample (g/100 g dry weight)} \times 100 \quad (1)$$

2.4. Color Evaluation

The surface color of uncooked and cooked samples was measured through a Minolta CR410 Chroma Meter (Konica Minolta Co., Tokyo, Japan). The CIELAB system color space (L*, a*, and b*) was considered. The D65 standard illuminant and a visual angle of 10 were used. Five readings were taken for each sample.

The total color difference (ΔE*) were calculated as follows:

$$\Delta E^*_{s-c} = [(L^*_s - L^*_c)^2 + (a^*_s - a^*_c)^2 + (b^*_s - b^*_c)^2]^{1/2} \quad (2)$$

where: s = annRS containing pasta and c = control. The ΔE* value > 3 indicates whether the color difference was perceivable by the human eye [23]. Before measuring, cooked pasta was carefully dried with absorbent paper.

2.5. Pasta Quality

The OCT was determined with the AACC-approved method 66-50 [24]. The cooking loss was determined by evaporating the cooking water to dryness at 105 °C (method 66-50) [24]. The water absorption capacity (WAC) was determined with the method AACC 66-50 [24]. Briefly, 25 g of pasta was cooked in 300 mL of boiling distilled water, rinsed in cold water, drained for 30 s, and weighed. The WAC was calculated as the relative weight increase after cooking.

2.6. Texture Properties

Texture characteristics (AACC method 66-50) [24] were conducted with a TA-XT2i Texture Analyser (Stable Micro Systems, UK) equipped with a 5 kg load cell. Cooked samples were dipped in cool water soon after cooking to stop the cooking process. Pasta firmness as maximum cutting force (AACC method 66-50) [24] was measured with a light knife blade (A/LKB) and a speed of 0.17 mm/s. From the force-time curve, the value of springiness was then derived. A pasta firmness/stickiness rig (HDP/PFS) at a compression speed of 0.5 mm/s and a compression force of 1 kg for 2 s was used to evaluate the stickiness (maximum peak force to separate the probe upon retraction from the sample's surface). Ten measurements for each sample were done.

2.7. Thermal Properties

The thermal properties of uncooked samples were studied through differential scanning calorimetry (DSC) (DSC8000, Perkin Elmer Inc., Waltham, MA, USA). Ground samples were weighed into steel pans, distilled water was added (1:3 *w*/*w* sample:water ratio), and the pans were sealed and left at room temperature. After 20 h, samples were heated from 25 to 170 °C at a rate of 10 °C/min. The onset temperature (T_0), the peak temperature (T_p), the conclusion temperature (T_c), and the gelatinization enthalpy (ΔH) were recorded using the software provided by the equipment. Results are expressed as the mean of 3 measurements for each sample.

2.8. In Vitro Starch Digestion of Gluten-Free Pasta

Samples (10 g) were cooked to optimum in 100 mL boiling water, drained up for 1 min, and directly analyzed. A 2-step (i.e., gastric and pancreatic phases) static in vitro starch digestion procedure was employed [25]. Cooked samples were passed through a meat mincer to mimic mastication, inserted in glass tubes, and hydrolyzed up to 180 min as detailed by Giuberti et al. [25]. Every 30 min up to 180 min liquid aliquots were taken for the measurement of the released glucose. This was done using a glucose oxidase kit (GODPOD 4058, Giesse Diagnostic snc, Rome, Italy). The area under the hydrolysis curve was measured and used to calculate the starch hydrolysis index (HI) with common white wheat bread as reference [25].

2.9. Sensory Analysis

The sensory profile of cooked to optimum macaroni pasta was evaluated by a 58-member panel recruited from students and staff of the Università Cattolica del Sacro Cuore (45% males and 55% females, 22–57 years old). Each member received 12 h of training prior to the test. Samples (750 g) were cooked to OCT in boiling salted water, and a cooked portion of 20 g was immediately offered to panelists. Each sample was labeled with three-digit random codes, and the order of presentation was balanced and randomized. Attributes included: color uniformity, appearance (regularity of shape, presence of deformation, cracks, and scratches), texture (hard at first chew), aroma, and taste. The test was carried out in one session, and members assigned the intensity of liking or disliking with a 9-point hedonic scale. Members were asked to comment on the overall acceptability using a 9-point hedonic scale (1–9). A score of 5 was considered as the limit of acceptability [26]. Water was provided between the evaluations. Each participant completed a written informed consent before the study.

2.10. Statistical Analyses

Data are presented as the mean values ± standard deviation of at least triplicate measurements. The comparison of means was conducted using the analysis of variance (One-way ANOVA) with a post hoc Tukey test at $p < 0.05$. The software IMB SPSS Statistics (Version 25) was used.

3. Results

3.1. Chemical Composition and Resistant Starch Content of Pasta

Irrespective of the annRS inclusion level, GF pasta samples were characterized by similar crude protein, crude lipid, and ash contents (Table 1).

Table 1. Chemical composition (g/100 g dry matter) and apparent resistant starch retention (aRSr, %) of gluten-free macaroni containing resistant starch (RS) from annealed white sorghum starch (annRS).

	Gluten-Free Pasta			
	Control [1]	**5-annRS [2]**	**10-annRS [3]**	**15-annRS [4]**
Moisture (g/100 g)	11.3 ± 0.33 [a]	10.9 ± 0.98 [a]	11.1 ± 0.08 [a]	11.8 ± 0.77 [a]
Total starch	87.6 ± 2.11 [a]	86.4± 1.91 [a]	84.3 ± 2.11 [b]	80.1 ± 2.13 [c]
Crude protein	8.0 ± 0.23 [a]	8.2 ± 0.55 [a]	7.9 ± 0.33 [a]	7.8 ± 0.77 [a]
Crude lipid	1.1 ± 0.11 [a]	1.3 ± 0.12 [a]	1.2 ± 0.09 [a]	1.2 ± 0.10 [a]
Ash	0.5 ± 0.01 [a]	0.4 ± 0.01 [a]	0.4 ± 0.02 [a]	0.3 ± 0.03 [a]
Dietary fiber	1.2 ± 0.12 [a]	4.0 ± 0.89 [b]	5.7 ± 0.72 [c]	9.2 ± 1.11 [d]
RS (uncooked)	0.7 ± 0.04 [a]	2.6 ± 0.12 [b]	4.3 ± 0.55 [c]	7.1 ± 0.82 [d]
RS (cooked to optimum)	0.04 ± 0.01 [a]	2.1 ± 0.03 [b]	3.6 ± 0.22 [c]	5.8 ± 0.59 [d]
aRSr (%)	5.7 ± 0.66	80.1 ± 2.33	83.2 ± 3.11	81.4 ± 3.27

Means in the same line with different superscript differed at $p < 0.05$. [1] Gluten-free macaroni prepared with 100% *w/w* rice flour. [2] Gluten-free macaroni prepared by mixing rice flour and annRS 95:5 *w/w*. [3] Gluten-free macaroni prepared by mixing rice flour and annRS 90:10 *w/w*. [4] Gluten-free macaroni prepared by mixing rice flour and annRS 85:15 *w/w*.

An increase in the TDF content was measured in GF pasta added with increasing levels of annRS, the highest value recorded for 15-annRS (i.e., 9.2 g/100 g DM, $p < 0.05$). The increase in the TDF following annRS inclusion is related to the analytical procedure employed, which measures the TDF by taking into account the RS and non-digestible oligosaccharides [27,28]. The current nutritional guidelines indicate that the definition of TDF includes carbohydrate polymers that are not hydrolyzed within the human small intestine. Accordingly, the RS, being classified as a functional fiber component, should be included [29]. Previous indications reported a low dietary fiber daily intake for individuals following a GF diet [30]. Accordingly, GF foods with high dietary fiber contents can be considered beneficial [31].

The RS is a functional dietary component that helps maintain metabolic and colonic health [32,33]. In the current study, the RS was measured prior to and after the cooking step to assess the thermal behavior of the selected RS-rich ingredient. The RS content of control pasta was 0.7 g/100 g DM, confirming previous findings on similar GF food products [11]. Besides, the annRS has proven effective in increasing the RS content, with the highest values recorded for 15-annRS pasta, both in the uncooked form and after the cooking step (i.e., 7.1 and 5.8 g/100 g DM, respectively). Accordingly, an aRSr of about 80% was calculated, irrespective of the level of annRS in the formulation (Table 1). Giuberti et al. [19] reported that annRS was characterized by higher thermal stability compared to the native white sorghum starch. This suggests that starch chain interactions formed during annealing are not disrupted during gelatinization, and this restricts the accessibility of the starch chains to the starch-hydrolyzing enzymes [33–35]. Findings agreed with those reported on GF biscuits made with increasing levels of annRS [21]. However, it is difficult to compare present findings with the literature because this is the first time in which the annRS was used in GF pasta formulation. Indeed, some studies suggested the addition of different RS-rich ingredients in wheat-based and GF pasta. Specifically, Gelencsér et al. [16] reported that the extrusion step did not cause a significant decrease in the RS content, but, on the contrary, greater RS loss was measured after cooking (on average −50%) in wheat pasta containing two different RS-rich ingredients (i.e., high amylose starch and a phosphate starch). Foschia et al. [11], using RS from high amylose maize, attributed the 30% loss in RS during pasta making without reporting data on the cooking step. In contrast, Aravind et al. [15] did not report changes comparing uncooked and cooked wheat pasta added with RS. Recently, Bresciani et al. [8] indicated that the pasta-making, but not the cooking step, significantly decreases the RS content in high amylose enriched pasta. Differences in the experimental conditions, RS sources, and applied food preparation process could explain the disagreement between studies.

3.2. Pasta Quality Evaluation and In Vitro Starch Digestion

Substitution of a part of rice flour with annRS resulted in color difference on uncooked samples, but only in marginal changes after cooking (Table 2). In particular, irrespective of the annRS addition level, cooked samples exhibited lower lightness and yellowness values than the uncooked counterparts. Results were consistent with Larrosa et al. [36], which reported a decrease in L* values of GF pasta after the cooking process. Moreover, in terms of total color difference, different ΔE* values were recorded only for uncooked 10- and 15-annRS samples, being >3 when compared to the control. After cooking, all annRS containing samples exhibited ΔE* values < 3, meaning that, as perceived by the human eye, the annRS containing samples were similar in color to the control.

Table 2. Quality parameters, texture analysis, and in vitro starch hydrolysis index of gluten-free macaroni containing resistant starch from annealed white sorghum starch (annRS).

	Gluten-Free Pasta			
	Control [1]	**5-annRS [2]**	**10-annRS [3]**	**15-annRS [4]**
Lightness L* (uncooked)	94.5 ± 0.16 c	93.2 ± 0.02 c	91.2 ± 0.13 b	89.1 ± 0.17 a
Redness a* (uncooked)	−0.3 ± 0.01 a	−0.4 ± 0.01 a	0.2 ± 0.02 b	0.4 ± 0.01 b
Yellowness b* (uncooked)	5.3 ± 0.05 a	5.0 ± 0.04 a	5.1 ± 0.72 a	5.2 ± 0.11 a
ΔE* (uncooked)	-	1.3	4.3	5.4
Lightness L* (cooked to optimum)	90.8 ± 0.22 a	89.8 ± 0.11 a	88.9 ± 0.20 a	88.4 ± 0.15 a
Redness a*(cooked to optimum)	−0.4 ± 0.01 a	−0.3 ±0.01 a	−0.3 ±0.02 a	−0.4 ± 0.01 a
Yellowness b* (cooked to optimum)	4.5 ± 0.05 b	3.3 ± 0.03 a	3.1 ± 0.02 a	2.8 ± 0.01 a
ΔE* (cooked to optimum)	-	1.6	2.4	2.9
Optimal cooking time (min)	9.3 ± 0.17 a	9.6 ± 0.22 a	10.6 ± 0.12 b	11.6 ± 0.34 b
Cooking loss (%)	12.2 ± 0.46 a	12.1 ± 0.33 a	10.4 ± 0.70 b	10.1 ± 0.27 b
Water absorption capacity (%)	101.3 ± 3.12 a	104.9 ± 2.22 b	107.2 ± 4.00 c	109.1 ± 3.51 c
Firmness (N)	1.6 ± 0.08 a	2.1 ± 0.10 b	2.3 ± 0.04 b	2.7 ± 0.11 c
Stickiness (N)	2.7 ± 0.21 b	2.6 ± 0.14 b	1.7 ± 0.15 a	1.4 ± 0.08 a
Springiness	0.44 ± 0.11 a	0.42 ± 0.09 a	0.44 ± 0.11 a	0.56 ± 0.12 b
In vitro starch hydrolysis index [5]	91.0 ± 3.12 d	83.2 ± 2.04 b	80.2 ± 3.01 b	73.1 ± 2.16 a

Means in the same line with different superscript differed at $p < 0.05$. [1] Gluten-free macaroni prepared with 100% w/w rice flour. [2] Gluten-free macaroni prepared by mixing rice flour and annRS 95:5 w/w. [3] Gluten-free macaroni prepared by mixing rice flour and annRS 90:10 w/w. [4] Gluten-free macaroni prepared by mixing rice flour and annRS 85:15 w/w. [5] Calculated using white wheat bread as reference (HI = 100 by definition).

Different optimal cooking time was recorded among samples, varying from 9.3 min for the control to 11.6 min for 15-annRS ($p < 0.05$). Foschia et al. [11] reported longer OCT in GF pasta supplemented with an RS-rich ingredient from high amylose maize. The cooking loss represents the percentage of DM lost in the cooking water. As reported in Table 2, a decrease in the cooking loss was observed when the level of annRS in the recipe accounts from 10 to 15% w/w, thus suggesting the formation of a structure with more resistance to disintegration on boiling. This is of interest since, in GF pasta, starch polymers are less efficiently entrapped in the matrix due to the lack of gluten, thus giving a final product with generally high cooking losses [37]. These findings are consistent with Foschia et al. [11] results in which the inclusion of RS (20% w/w) to GF pasta led to a decrease in the cooking loss of about 30%. Lower values of cooking loss are considered desirable because they indicate a lower solubility of starch and a greater cooking tolerance [38]. Concerning the WAC values, higher values were recorded as the level of annRS increased in the recipe ($p < 0.05$). According to Sozer et al. [39], a longer cooking time corresponds to an increase in water absorption since more water can diffuse and interact with starch. In addition, the greater WAC of samples containing annRS can be related to the inherent characteristics of the selected RS-rich ingredient, characterized by a high WAC value and a greater ability to expose hydrophilic groups to bind water molecules [19].

Firmness, stickiness, and springiness are important attributes used to evaluate pasta quality [40]. The addition of annRS significantly increased the firmness (as maximum cutting force) of the cooked pasta, with values ranging from 1.6 N to 2.7 N for control and 15-annRS pasta, respectively ($p < 0.05$). This suggests the presence of a more compact structure following annRS inclusion in the recipe. Similar results were reported by Foschia et al. [11] in GF pasta enriched with 10–20% of RS, while Sozer et al. [39] using green banana starch as a source of RS did not report a significant effect on firmness. According to Marti and Pagani [37], the inclusion of different starch types at different levels in pasta formulation can contribute to modify the firmness of the final product to different extents due to the inherent starch characteristic, the specific starch network created during the pasta making and interactions occurring on cooking. Substitution of rice flour with annRS led to changes also in the stickiness parameter (Table 2). According to the literature, cooked pasta should have minimal stickiness values [41]. In this work, stickiness values decreased as the level of annRS increased, the lowest value recorded for 15-annRS (i.e., 1.4 N: $p < 0.05$). Results are consistent with Aravind et al. [15] and Foschia et al. [11], which reported lower stickiness values in RS-enriched pasta. According to the authors, the macromolecular reorganization induced by the RS addition could prevent excessive leaching of starch during the cooking process, preventing, in this way, stickiness and excessive cooking losses. The springiness indicates the ability of pasta to recuperate its original shape after compression. In general, GF pasta generally lacks elasticity compared to wheat of durum pasta [37]. The substitution of rice flour with annRS led to changes in springiness only at the highest level of ann-RS inclusion in the recipe ($p < 0.05$). In addition, recorded springiness values appeared in line with literature data for GF pasta but still inferior compared to gluten-containing counterparts [15,17,37].

The in vitro starch digestion curves are presented in Figure 1.

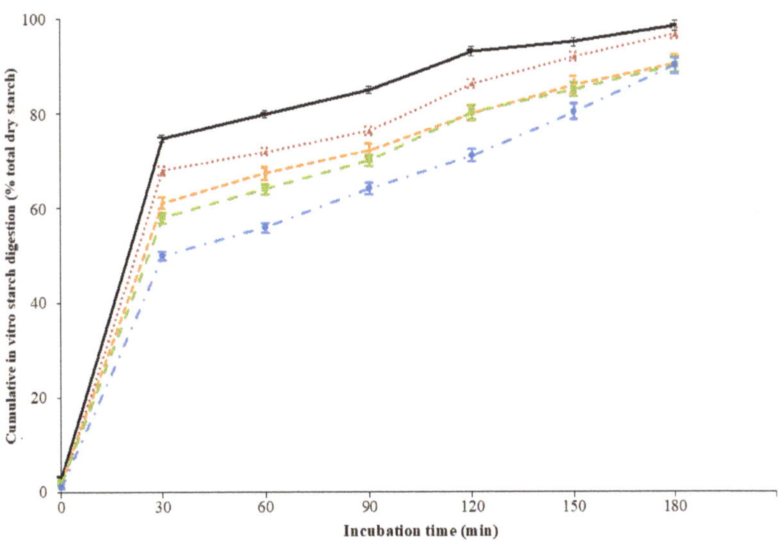

Figure 1. In vitro starch digestion curves of gluten-free macaroni containing resistant starch from annealed white sorghum starch (annRS). Control: gluten-free macaroni prepared with 100% *w*/*w* rice flour (red line); 5-annRS: gluten-free macaroni prepared by mixing rice flour and annRS 95:5 *w*/*w* (orange line); 10-annRS: gluten-free macaroni prepared by mixing rice flour and annRS 90:10 *w*/*w* (green line); 15-annRS: gluten-free macaroni prepared by mixing rice flour and annRS 85:15 *w*/*w* (blue line). White wheat bread is used as a reference (black line). Before analyses, pasta samples were cooked to optimal cooking time.

The in vitro starch digestion curve of white wheat bread was in line with previous findings [42]. In addition, the higher RS content of pasta sample following annRS inclusion level reflected in a different extent of the in vitro starch digestibility of the cooked samples. The starch HI decreased significantly ($p < 0.05$) as the level of substitution of annRS increased, with values ranging from 91 for control pasta to about 73 for 15-annRS. This can be related to the lower susceptibility of annRS to enzymatic digestion. In cereal products, the RS fraction is not digestible neither in vitro nor in vivo; consequently, RS does not contribute to the release of glucose during the enzyme hydrolysis, which leads to a decrease in the starch HI [33,43]. Similar results have been reported for GF biscuits [21]. In addition, the possible role of the product's hardness could also partially preserve the starch's granular structural integrity during cooking and/or modulate the in vitro accessibility of enzymes to starch. This might have contributed towards the reduction in the starch HI of the samples, in line with previous indications [44,45].

3.3. Thermal Properties

The thermal properties of GF pasta samples are presented in Table 3. Control pasta was characterized by T_0, T_p, T_c, and ΔH mean values of 60.3 °C, 68.2 °C, 74.4 °C, and 3.2 J/g. Marti et al. [38] reported that 100% rice pasta made with a conventional extrusion process exhibited a peak in the range of 55.4–72.5 °C, in line with the current findings. The data obtained by DCS suggested that 10-annRS and 15-annRS samples required higher temperature values for melting (from 68.3 to 86.2 °C and from 73.1 to 89.1 °C, respectively) with respect to 5-annRS and control pasta, thus resulting in a GF pasta more stable during heating. In addition, both 10-annRS and 15-annRS pasta required more energy for gelatinization (on average 5 J/g) than the other samples. Taken together, present DSC findings indicated greater thermal stability of GF pasta formulated by replacing rice flour with at least 10 g/100 g (*w*/*w*) of ann-RS. These results are consistent with the pasta behavior on cooking: the strong network obtained following annRS addition at greater inclusion level in the recipe may contribute to explain the lower cooking loss value reported for 10-annRS and 15-ann-RS pasta (Table 2), in line with previous findings [38]. In addition, reported differences in the starch gelatinization properties among samples might be related to the thermal properties of annRS and to possible differences in the starch organization/architecture following annRS inclusion during the pasta-making process. In particular, Giuberti et al. [19] reported that annealed white sorghum starch was characterized by the greatest ΔH values (i.e., 14.6 J/g), along with the greater gelatinization transition temperatures when compared to the native counterpart.

Table 3. Thermal properties of gluten-free macaroni containing resistant starch from annealed white sorghum starch (annRS).

	Gluten-Free Pasta			
	Control [1]	**5-annRS** [2]	**10-annRS** [3]	**15-annRS** [4]
Onset temperature T_0 (°C)	60.3 ± 2.16 ^{a}	62.1 ± 1.43 ^{a}	68.3 ± 1.93 ^{b}	73.1 ± 2.33 ^{c}
Peak temperature T_p (°C)	68.2 ± 1.55 ^{a}	70.3 ± 0.94 ^{a}	81.2 ± 2.02 ^{b}	84.7 ± 1.01 ^{c}
Conclusion temperature T_c (°C)	74.4 ± 1.02 ^{a}	77.3 ± 0.04 ^{a}	86.2 ± 0.82 ^{b}	89.1 ± 0.61 ^{b}
Gelatinization enthalpy ΔH (J/g)	3.2 ± 0.17 ^{a}	3.4 ± 0.34 ^{a}	4.8 ± 0.23 ^{b}	5.2 ± 0.14 ^{b}

Means in the same line with different superscript differed at $p < 0.05$. [1] Gluten-free macaroni prepared with 100% *w*/*w* rice flour. [2] Gluten-free macaroni prepared by mixing rice flour and annRS 95:5 *w*/*w*. [3] Gluten-free macaroni prepared by mixing rice flour and annRS 90:10 *w*/*w*. [4] Gluten-free macaroni prepared by mixing rice flour and annRS 85:15 *w*/*w*.

3.4. Sensory Analysis

The mean values for each sensorial attribute of control and annRS enriched pasta samples are presented in Table 4.

Table 4. Average sensory scores of gluten-free macaroni containing resistant starch from annealed white sorghum starch (annRS).

	Gluten-Free Pasta			
	Control [1]	**5-annRS [2]**	**10-annRS [3]**	**15-annRS [4]**
Color	5.4 ± 0.56 [a]	5.5 ± 0.12 [a]	5.3 ± 0.34 [a]	5.3 ± 0.45 [a]
Appearance	6.1 ± 0.32 [a]	6.1 ± 0.54 [a]	5.8 ± 0.61 [a]	5.3 ± 0.31 [b]
Texture	4.6 ± 0.15 [a]	4.8 ± 0.32 [a]	5.2 ± 0.72 [b]	5.9 ± 0.22 [c]
Aroma	5.4 ± 0.32 [a]	5.5 ± 0.43 [a]	5.3 ± 0.11 [a]	5.3 ± 0.27 [a]
Taste	5.0 ± 0.41 [a]	5.0 ± 0.33 [a]	5.1 ± 0.31 [a]	5.1 ± 0.66[a]
Overall acceptance	6.3 ± 0.32 [b]	6.1 ± 0.14 [b]	5.4 ± 3.01 [a]	5.3 ± 2.16 [a]

Means in the same line with different superscript differed at $p < 0.05$. [1] Gluten-free macaroni prepared with 100% *w*/*w* rice flour. [2] Gluten-free macaroni prepared by mixing rice flour and annRS 95:5 *w*/*w*. [3] Gluten-free macaroni prepared by mixing rice flour and annRS 90:10 *w*/*w*. [4] Gluten-free macaroni prepared by mixing rice flour and annRS 85:15 *w*/*w*.

No significant difference was observed among samples in color, aroma, or taste, with average values of 5.4, 4.6, and 5.1, respectively, thus indicating that the type of RS, along with its relative amount in the recipe, did not cause changes in these parameters according to the sensory panel. This is probably related to the neutral flavor of the annRS ingredient [21–43]. These findings appear consistent with Gelencsér et al. [46], which reported no differences between pasta enriched with RS and control wheat-based pasta. Sensory scores for color attributes appeared in line with the instrumental values in which color differences between cooked control pasta and annRS enriched pasta were not detected. In addition, the texture of 10- and 15-annRS were relatively more appreciated by panelists with respect to the other pasta samples, thus confirming the effect on texture measured by the instrumental analysis. The appearance attribute showed the lowest score for the 15-annRS sample (5.3; $p < 0.05$), and a significant decrease in the overall acceptance was measured as the level of annRS increased in the formulation. However, all samples resulted in a score higher than 5 for the overall acceptance value, which is considered as the limit of acceptability [26]. Taken together, present data shows that the addition of annRS to pasta up to 15 g/100g *w*/*w* has a minimal effect on the sensory attributes, in line with previous findings [15,46].

4. Conclusions

The annRS has potential application as a value-added ingredient to produce GF pasta with high RS content and lower in vitro starch digestion with respect to 100% rice counterpart. The substitution of common rice flour with 15 g/100 g *w*/*w* of annRS also allows using the "high in fiber" claim [47]. Blending rice flour with increasing levels of annRS resulted in longer optimal cooking time, lower cooking losses, along with positive changes in texture and stickiness, thus suggesting the formation of a structure with more resistance to boiling. However, the lightness of uncooked pasta decreased as the level of annRS increased in the recipe, which may potentially reduce the attractiveness of the new formulated GF pasta to consumers. Sensory attributes were only marginal affected by the annRS inclusion. Present findings underline the suitability of this RS ingredient in GF pasta production up to 15 g/100 g *w*/*w*. Further studies to assess the in vivo digestibility and potential health benefits are desirable.

Author Contributions: Conceptualization, M.C. and A.G.; validation, G.G.; formal analysis, M.C. and A.G.; investigation, M.C. and A.G.; resources, G.G.; data curation, G.G. and A.B.; writing—original draft preparation, M.C. and A.G.; writing—review and editing, G.S. and G.G.; visualization, G.G.; supervision, G.G.; project administration, G.G.; funding acquisition, G.S. and G.G. All authors have read and agreed to the published version of the manuscript.

Funding: This research received no external funding.

Institutional Review Board Statement: The study was conducted according to the guidelines of the Declaration of Helsinki, and approved by the Ethics Committee of Policlinico Gemelli Hospital (authorization number 5289/15).

Informed Consent Statement: Informed consent was obtained from all subjects involved in the sensory study.

Acknowledgments: The research was supported by the 5 × 1000–Dichiarazione 2019–Università Cattolica del Sacro Cuore, Largo Agostino Gemelli, 1, 20123 Milano, MI, Italy.

Conflicts of Interest: The authors declare no conflict of interest.

References

1. Padalino, L.; Conte, A.; Del Nobile, M.A. Overview on the general approaches to improve gluten-free pasta and bread. *Foods* **2016**, *5*, 87. [CrossRef] [PubMed]
2. Giuberti, G.; Gallo, A. Reducing the glycaemic index and increasing the slowly digestible starch content in gluten-free cereal-based foods: A review. *Int. J. Food Sci. Technol.* **2018**, *53*, 50–60. [CrossRef]
3. Fry, L.; Madden, A.M.; Fallaize, R. An investigation into the nutritional composition and cost of gluten-free versus regular food products in the UK. *J. Hum. Nutr. Diet.* **2018**, *31*, 108–120. [CrossRef] [PubMed]
4. Woomer, J.S.; Adedeji, A.A. Current applications of gluten-free grains–a review. *Crit. Rev. Food Sci. Nutr.* **2021**, *61*, 14–24 [CrossRef]
5. Giuberti, G.; Marti, A.; Fortunati, P.; Gallo, A. Gluten free rice cookies with resistant starch ingredients from modified waxy rice starches: Nutritional aspects and textural characteristics. *J. Cereal Sci.* **2017**, *76*, 157–164. [CrossRef]
6. Allen, B.; Orfila, C. The availability and nutritional adequacy of gluten-free bread and pasta. *Nutrients* **2018**, *10*, 1370. [CrossRef]
7. El Khoury, D.; Balfour-Ducharme, S.; Joye, I.J. A review on the gluten-free diet: Technological and nutritional challenges. *Nutrients* **2018**, *10*, 1410. [CrossRef]
8. Bresciani, A.; Giordano, D.; Vanara, F.; Blandino, M.; Marti, A. High-amylose corn in gluten-free pasta: Strategies to deliver nutritional benefits ensuring the overall quality. *Food Chem.* **2021**, *353*, 129489. [CrossRef] [PubMed]
9. Arribas, C.; Cabellos, B.; Cuadrado, C.; Guillamón, E.; Pedrosa, M.M. Cooking Effect on the Bioactive Compounds, Texture, and Color Properties of Cold-Extruded Rice/Bean-Based Pasta Supplemented with Whole Carob Fruit. *Foods* **2020**, *9*, 415. [CrossRef]
10. Flores-Silva, P.C.; Rodriguez-Ambriz, S.L.; Bello-Pérez, L.A. Gluten-free snacks using plantain–chickpea and maize blend: Chemical composition, starch digestibility, and predicted glycemic index. *J. Food Sci.* **2015**, *80*, C961–C966. [CrossRef]
11. Foschia, M.; Beraldo, P.; Peressini, D. Evaluation of the physicochemical properties of gluten-free pasta enriched with resistant starch. *J. Sci. Food Agric.* **2017**, *97*, 572–577. [CrossRef]
12. Raungrusmee, S.; Shrestha, S.; Sadiq, M.B.; Anal, A.K. Influence of resistant starch, xanthan gum, inulin and defatted rice bran on the physicochemical, functional and sensory properties of low glycemic gluten-free noodles. *LWT Food Sci. Technol.* **2020**, *126*, 109279. [CrossRef]
13. Birt, D.F.; Boylston, T.; Hendrich, S.; Jane, J.L.; Hollis, J.; Li, L.; McClelland, J.; Moore, S.; Phillips, G.J.; Rowling, M.; et al. Resistant starch: Promise for improving human health. *Adv. Nutr.* **2013**, *4*, 587–601. [CrossRef] [PubMed]
14. Miketinas, D.C.; Shankar, K.; Maiya, M.; Patterson, M.A. Usual dietary intake of resistant starch in US adults from NHANES 2015–2016. *J. Nutr.* **2020**, *150*, 2738–2747. [CrossRef]
15. Aravind, N.; Sissons, M.; Fellows, C.M.; Blazek, J.; Gilbert, E.P. Optimisation of resistant starch II and III levels in durum wheat pasta to reduce in vitro digestibility while maintaining processing and sensory characteristics. *Food Chem.* **2013**, *136*, 1100–1109. [CrossRef]
16. Gelencsér, T.; Gál, V.; Salgó, A. Effects of applied process on the in vitro digestibility and resistant starch content of pasta products. *Food Bioprocess Technol.* **2010**, *3*, 491–497. [CrossRef]
17. Vernaza, M.G.; Biasutti, E.; Schmiele, M.; Jaekel, L.Z.; Bannwart, A.; Chang, Y.K. Effect of supplementation of wheat flour with resistant starch and monoglycerides in pasta dried at high temperatures. *Int. J. Food Sci. Technol.* **2012**, *47*, 1302–1312. [CrossRef]
18. Masato, O.; Kentaro, M.; Tatsuro, M.; Akio, F.; Yukako, H.; Yasuki, M. Effects of drying temperature on the properties of starch in pasta. *LWT Food Sci. Technol.* **2021**, *145*, 111171. [CrossRef]
19. Giuberti, G.; Marti, A.; Gallo, A.; Grassi, S.; Spigno, G. Resistant starch from isolated white sorghum starch: Functional and physicochemical properties and resistant starch retention after cooking. A comparative study. *Starch Stärke* **2019**, *71*, 1800194. [CrossRef]
20. Jiang, F.; Du, C.; Jiang, W.; Wang, L.; Du, S.K. The preparation, formation, fermentability, and applications of resistant starch. *Int. J. Biol. Macromol.* **2020**, *150*, 1155–1161. [CrossRef]
21. Cervini, M.; Frustace, A.; Duserm Garrido, G.; Rocchetti, G.; Giuberti, G. Nutritional, physical and sensory characteristics of gluten-free biscuits incorporated with a novel resistant starch ingredient. *Helyon* **2021**, *7*, e06562. [CrossRef] [PubMed]
22. Association of Official Analytical Chemists. *Official Methods of Analysis*, 17th ed.; Association of Official Analytical Chemists: Gaithersburg, MD, USA, 2000.

23. Laguna, L.; Salvador, A.; Sanz, T.; Fiszman, S.M. Performance of a resistant starch rich ingredient in the baking and eating quality of short-dough biscuits. *LWT Food Sci. Technol.* **2011**, *44*, 737–746. [CrossRef]
24. American Association of Cereal Chemists. *Approved Methods of the AACC*, 10th ed.; American Association of Cereal Chemists: St. Paul, MN, USA, 2000; pp. 66–150.
25. Giuberti, G.; Gallo, A.; Cerioli, C.; Fortunati, P.; Masoero, F. Cooking quality and starch digestibility of gluten free pasta using new bean flour. *Food Chem.* **2015**, *175*, 43–49. [CrossRef]
26. Meilgaard, M.C.; Carr, B.T.; Civille, G.V. *Sensory Evaluation Techniques*; CRC Press: Boca Raton, FL, USA, 2006.
27. McCleary, B.V. An integrated procedure for the measurement of total dietary fibre (including resistant starch), non-digestible oligosaccharides and available carbohydrates. *Anal. Bioanal. Chem.* **2007**, *389*, 291–308. [CrossRef]
28. McCleary, B.V.; De Vries, J.W.; Rader, J.I.; Cohen, G.; Prosky, L.; Mugford, D.C.; Champ, M.; Okuma, K. Determination of total dietary fiber (CODEX definition) by enzymatic-gravimetric method and liquid chromatography: Collaborative study. *J. AOAC Int.* **2010**, *93*, 221–233. [CrossRef]
29. Betteridge, V. Dietary fibre: An evolving definition? *Nutr. Bull.* **2009**, *34*, 122–125.
30. Taetzsch, A.; Das, S.K.; Brown, C.; Krauss, A.; Silver, R.E.; Roberts, S.B. Are gluten-free diets more nutritious? An evaluation of self-selected and recommended gluten-free and gluten-containing dietary patterns. *Nutrients* **2018**, *10*, 1881. [CrossRef]
31. Lamacchia, C.; Camarca, A.; Picascia, S.; Di Luccia, A.; Gianfrani, C. Cereal-based gluten-free food: How to reconcile nutritional and technological properties of wheat proteins with safety for celiac disease patients. *Nutrients* **2014**, *6*, 575–590. [CrossRef]
32. Halajzadeh, J.; Milajerdi, A.; Reiner, Ž.; Amirani, E.; Kolahdooz, F.; Barekat, M.; Mirzaei, H.; Mirhashemi, S.M.; Asemi, Z. Effects of resistant starch on glycemic control, serum lipoproteins and systemic inflammation in patients with metabolic syndrome and related disorders: A systematic review and meta-analysis of randomized controlled clinical trials. *Crit. Rev. Food Sci. Nutr.* **2020**, *60*, 3172–3184. [CrossRef]
33. Tian, S.; Sun, Y. Influencing factor of resistant starch formation and application in cereal products: A review. *Int. J. Biol. Macromol.* **2020**, *149*, 424–431. [CrossRef] [PubMed]
34. Chung, H.J.; Liu, Q.; Hoover, R. Impact of annealing and heat-moisture treatment on rapidly digestible, slowly digestible and resistant starch levels in native and gelatinized corn, pea and lentil starches. *Carbohydr. Polym.* **2009**, *75*, 436–447. [CrossRef]
35. Jayakody, L.; Hoover, R. Effect of annealing on the molecular structure and physicochemical properties of starches from different botanical origins–A review. *Carbohydr. Polym.* **2008**, *74*, 691–703. [CrossRef]
36. Larrosa, V.; Lorenzo, G.; Zaritzky, N.; Califano, A. Improvement of the texture and quality of cooked gluten-free pasta. *LWT Food Sci. Technol.* **2016**, *70*, 96–103. [CrossRef]
37. Marti, A.; Pagani, M.A. What can play the role of gluten in gluten free pasta? *Trends Food Sci. Technol.* **2013**, *31*, 63–71. [CrossRef]
38. Marti, A.; Seetharaman, K.; Pagani, M.A. Rice-based pasta: A comparison between conventional pasta-making and extrusion-cooking. *J. Cereal Sci.* **2010**, *52*, 404–409. [CrossRef]
39. Sozer, N.; Dalgıc, A.C.; Kaya, A. Thermal, textural and cooking properties of spaghetti enriched with resistant starch. *J. Food Eng.* **2007**, *81*, 476–484. [CrossRef]
40. Palavecino, P.M.; Curti, M.I.; Bustos, M.C.; Pencil, M.C.; Ribotta, P.D. Sorghum Pasta and Noodles: Technological and Nutritional Aspects. *Plant. Foods Hum. Nutr.* **2020**, *75*, 326–336. [CrossRef]
41. Lucisano, M.; Cappa, C.; Fongaro, L.; Mariotti, M. Characterization of gluten-free pasta through conventional and innovative methods: Evaluation of the cooking behaviour. *J. Cereal Sci.* **2012**, *56*, 667–675. [CrossRef]
42. Jia, M.; Yu, Q.; Chen, J.; He, Z.; Chen, Y.; Xie, J.; Liu, X.; Nie, S.; Xie, M. Physical quality and in vitro starch digestibility of biscuits as affected by addition of soluble dietary fiber from defatted rice bran. *Food Hydrocoll.* **2020**, *99*, 105349. [CrossRef]
43. Sajilata, M.G.; Singhal, R.S.; Kulkarni, P.R. Resistant starch—A review. *Compr. Rev. Food Sci. Food Saf.* **2006**, *5*, 1–17. [CrossRef]
44. Pellegrini, N.; Vittadini, E.; Fogliano, V. Designing food structure to slow down digestion in starch-rich products. *Curr. Opin. Food Sci.* **2020**, *32*, 50–57. [CrossRef]
45. Sparvoli, F.; Laureati, M.; Pilu, R.; Pagliarini, E.; Toschi, I.; Giuberti, G.; Fortunati, P.; Daminati, M.G.; Cominelli, E.; Bollini, R. Exploitation of common bean flours with low antinutrient content for making nutritionally enhanced biscuits. *Front. Plant. Sci.* **2016**, *7*, 928. [CrossRef] [PubMed]
46. Gelencsér, T.; Gál, V.; Hódsági, M.; Salgó, A. Evaluation of quality and digestibility characteristics of resistant starch-enriched pasta. *Food Bioprocess. Technol.* **2008**, *1*, 171–179. [CrossRef]
47. European Commission. Commission Regulation (EC) No. 1924/2006, of 20 December 2006 on nutrition and health claims made on foods. *Off. J. Eur. Union* **2006**, *404*, 9–25.

foods

Article

Traditional and Non-Conventional Pasta-Making Processes: Effect on In Vitro Starch Digestibility

Rossella Dodi [1], Letizia Bresciani [1], Beatrice Biasini [1], Marta Cossu [1], Francesca Scazzina [1], Federica Taddei [2], Maria Grazia D'Egidio [2], Margherita Dall'Asta [3,*] and Daniela Martini [4]

1 Department of Food and Drug, Human Nutrition Unit, University of Parma, 43125 Parma, Italy; rossella.dodi@unipr.it (R.D.); letizia.bresciani@unipr.it (L.B.); beatrice.biasini@unipr.it (B.B.); cossu.marta@gmail.com (M.C.); francesca.scazzina@unipr.it (F.S.)
2 CREA Research Centre for Engineering and Agro-Food Processing, 00189 Rome, Italy; federica.taddei@crea.gov.it (F.T.); mariagrazia.degidio@gmail.com (M.G.D.)
3 Department of Animal Science, Food and Nutrition, Università Cattolica del Sacro Cuore, 29122 Piacenza, Italy
4 Department of Food, Environmental and Nutritional Sciences (DeFENS), Università Degli Studi di Milano, 20133 Milan, Italy; daniela.martini@unimi.it
* Correspondence: margherita.dallasta@unicatt.it

Abstract: Pasta is a carbohydrate-rich food with a low glycemic index (GI) and is one of the main sources of slowly digestible starch (SDS). The presence of bran fractions (BFs) in pasta may enhance its health potential, owing to the content of fiber, micronutrients, and bioactive compounds; however, at the same time, BF may affect starch digestibility. In this study, the bioaccessibility of starch in pasta made with BF-enriched semolina (BF pasta), or only with micronized debranned kernel (DK pasta), and a control pasta made with traditional semolina was evaluated by applying two different in vitro models. The control pasta showed a percentage of SDS about four-fold higher than that of the BF pasta and 1.5-fold higher than that of the DK pasta ($p < 0.05$). The amount of starch released during simulated gastrointestinal digestion was slightly lower, but not significantly different, for the control pasta than for both the BF and DK pasta. These results suggest that the presence of a higher amount of dietary fiber in BF pasta can affect the structure of the food matrix, interfering with the formation of the gluten network, water absorption, and starch granule accessibility, while micronization could enhance starch digestibility due to starch gelatinization. These findings emphasize the need to optimize the process for producing fiber-rich pasta without affecting its low starch digestibility and, consequently, its GI.

Keywords: slowly digestible starch; carbohydrates; fiber; in vitro digestion; micronization

Citation: Dodi, R.; Bresciani, L.; Biasini, B.; Cossu, M.; Scazzina, F.; Taddei, F.; D'Egidio, M.G.; Dall'Asta, M.; Martini, D. Traditional and Non-Conventional Pasta-Making Processes: Effect on In Vitro Starch Digestibility. *Foods* **2021**, *10*, 921. https://doi.org/10.3390/foods10050921

Academic Editors: Joan M. King

Received: 21 March 2021
Accepted: 20 April 2021
Published: 22 April 2021

Publisher's Note: MDPI stays neutral with regard to jurisdictional claims in published maps and institutional affiliations.

1. Introduction

Pasta is one of the staple foods of the Mediterranean diet, and it is widely produced and consumed all over the world [1]. In 2019, in Italy—where pasta represents one of the key foods of the gastronomic tradition—consumption reached about 23 kg/per capita per year [1]. Depending on the total energy intake level, Italian dietary guidelines suggest the daily consumption of 3.5 to 6 portions of carbohydrate-rich foods, such as pasta, rice, and other cereals or cereal-derived products [2], which contribute to the intake of complex carbohydrates both in Italy and in many other countries, even those outside the Mediterranean basin [1,3]. Due to its wide consumption, pasta contributes to covering the recommended 45–60% of the total daily energy intake from carbohydrates and, if consumed as wholegrain pasta, the suggested dietary target for fiber (at least 25 g per day) [4].

Besides available starch, pasta—especially when consumed as whole-wheat pasta—is indeed a good carrier of fiber, micronutrients, and bioactive compounds [5,6]. In particular, whole-wheat semolina is rich in fiber, mainly insoluble fiber (e.g., cellulose, arabinoxylans), which is well-recognized for its beneficial role in bowel health and many other health outcomes, including a positive role in metabolic health and, thus, in the prevention of several chronic diseases [6–8]. Despite these beneficial effects associated with wholegrain products, the consumption of whole-wheat pasta in Italy still seems limited [9], even though recent data showed a substantial increase in the past decade of launches on the market and the consumption of wholegrain products [10]. Potential barriers to the consumption of wholegrain foods include personal, product-specific, and external factors, such as sensory aspects and dietary habits [11,12]. In particular, inadequate knowledge of the positive effect of wholegrain consumption on chronic disease risk reduction often leads consumers to prefer refined grain products [13,14].

The food structure of pasta is the result of changes occurring in its main components, namely, starch and protein, throughout the technological production process [15], which, in turn, influences the nutritional quality of this product. Indeed, pasta processing leads to an increase in the slowly digestible starch (SDS) fraction in the final product, which elicits a lower post-prandial glycemic response after consumption compared to that from other cereal-based products, such as rice and bread [16]. To date, this is one of the main evidenced reasons behind the beneficial effect of pasta consumption [17–19]. The use of innovative technological processes has been proposed as a strategy to produce cereal-based products such as pasta that preserve the natural health properties of grains while limiting the negative aspects related to the use of wholegrain products [20]. Among these strategies, the use of debranning has been proposed. Debranning consists of pearling and pealing of the kernels that can be used to obtain both selected bran fractions (BFs) and debranned kernels (DKs) [21,22]. Different studies have demonstrated that debranning is a pretreatment potentially able to improve milling yields [23]; moreover, the use of debranning products allows for the making of pasta with a high content of dietary fibers, vitamins, and phenolic compounds and minimal effects on sensory properties, using only the natural endowment of durum wheat [21,24,25]. Additionally, preprocessing could reduce the total microbial contamination and the content of mycotoxins or heavy metals in flour, which, in turn, affect its safety and quality [23,26–28].

However, the modification of process parameters may also affect the pasta structure and potentially change the digestibility of the starch and protein fractions [29,30]. For instance, the addition of BF or the use of a different process in pasta-making can affect starch digestibility. In fact, the presence of bran within the wholemeal pasta matrix may physically interfere with the gluten matrix, making the structure highly porous, which, in turn, increases the accessibility of the starch granules to α-amylase during digestion [15]. Therefore, investigating the starch digestibility in vitro is of physiologic relevance and represents a useful approach for predicting the in vivo bioavailability of carbohydrates contained in pasta and the glycemic index (GI) [31,32].

Thus, because the modification of process parameters may influence the digestibility of starch, influencing both the accessibility to the digestive enzymes and, consequently, the glycemic response in vivo, the aim of the present study was to investigate the in vitro digestibility of starch in pasta produced by using BFs or DKs and to compare it with that of a traditional pasta made with semolina.

2. Materials and Methods

2.1. Chemicals

Reagents and enzymes were purchased from Sigma-Aldrich (St. Louis, MO, USA), unless otherwise noted. The enzymes used are reported as follows: pepsin (EC number 3.4.23.1), pancreatin (EC number 232-468-9), guar (EC number 232-536-8), invertase (EC number 3.2.1.26), amyloglucosidase (AMG) from E-AMGDF Megazyme kit (Wicklow, Ireland).

2.2. Debranning and Traditional Milling Processes

The Italian *Triticum durum* wheat Normanno, a widely used Italian durum wheat cultivar, was used as raw materials for the whole experiment. In detail, an aliquot of kernels was debranned three sequential times for about 90 s each, using a pilot plant (NAMAD, Rome, Italy), to obtain three bran fractions (BFs 1, BFs 2, BFs 3) and aliquots of the resulting kernels (DKs 1, DKs 2, DKs 3), corresponding to ranges of debranning levels (DLs) of 0–2.80%, 2.81–5.10%, and 5.11–8.00%, respectively. An additional aliquot of the non-debranned kernels of the same cultivar was traditionally milled in a pilot plant (Buhler MLU 202, Uzwil, Switzerland) to obtain semolina.

2.3. Pasta Samples: Preparation and Cooking

Some of the debranning products described in Section 2.1 (i.e., BF2 and DK1) and semolina were used to produce two different pasta samples: (i) BF pasta, produced by enriching semolina with BFs 2 (BFs 2: semolina ratio of 30:100 *w/w*); and (ii) DK pasta, produced by using only micronized DKs 1 that still include BF2 used for the BF pasta. DK 1 was micronized using a KMX-500 micronizer (Separmicrosystem S.a.S, Brescia, Italy). The third pasta type (control pasta, CTRL), used as a reference, was made by using only semolina obtained as described in Section 2.2. The three samples were processed into spaghetti by an experimental press (NAMAD, Rome, Italy) and were dried using an experimental drier (AFREM, Lyon, France) for 20 h, applying a low-temperature drying process (50 °C). The pasta-making process was repeated twice. A cooking test was performed by adding 100 g of dried pasta to 1 L of boiling tap water with a standard cooking time of 13 min. Exhaustive information on the technological process and nutritional composition of the pasta samples was reported in a previous study [24].

2.4. Available Starch Determination

Determination of the available starch (Av starch) was carried out using the AOAC Method 2002.02, AACC Method 32-40.01 (Megazyme assay kit, K-RSTAR). The available starch analysis was performed according to the manufacturer's instruction, with slight modifications.

Briefly, 150 mg of cooked and minced pasta was weighed and incubated in a Dubnoff bath (ISCO, Milan, Italy) with pancreatic α-amylase and AMG for 16 h at 37 °C and 180 strokes/min. After the incubation, 4 mL of 100% ethanol was added to each sample and centrifuged at 3000 rpm for 10 min. After the centrifugation, the supernatant was transferred into a 100 mL matrass. The tubes containing the pellet were washed three times with 50% aqueous ethanol, dissolved in 2 mL of KOH (2 M), and put on ice. After 20 min, 8 mL of sodium acetate buffer (1.2 M, pH 3.8) was added into the samples, and after the addition of 100 µL of AMG (3300 U/mL), the samples were incubated at 50 °C for 30 min. The samples were centrifuged for 10 min at 3000 rpm and stored at −20 °C until the resistant starch analysis.

A volume of 100 mL of sodium acetate buffer (100 mM, pH 4.5) was added into the matrass containing the supernatant to adjust the total volume. Quantities of 200 µL of the solution were transferred into tubes, and 20 µL of AMG (300 U/mL) was added. The tubes were incubated for 20 min at 50 °C, and the available starch was quantified in the supernatant of the samples by means of an automatic glucose analyzer (model 2900, Yellow Springs Instrument Company, Yellow Springs, OH, USA). The analyses were performed in triplicate for each sample.

2.5. In Vitro Starch Digestibility

In vitro digestion of the pasta was performed following the method of Brighenti and colleagues [32], with some modifications [33]. Samples were cooked and extruded through 7 mm holes of a hand-operated mincer (Sirius, Karl Krüger). Briefly, 8 g samples were weighed and suspended in 5 mL of preheated (37 °C) 20 mM sodium phosphate buffer (pH 6.9, 10 mM NaCl) and 25 mL of preheated (37 °C) 0.9% NaCl with 1.5 mL of human saliva. Saliva was collected from three non-smoking adult donors after careful tooth brushing and abstinence from food and drink for at least 1 h prior to the experiment. After 2 min incubation in a shaking water bath (SW23, Julabo, Milan, Italy) at 37 °C and 160 strokes/min, the pH was adjusted to 2–2.5 using 5 M HCl. One milliliter of a solution of 0.9% NaCl dissolved porcine pepsin (2500 U/mL pepsin, was then added to each sample in order to mimic the gastric phase. The mixtures were incubated in a shaking water bath at 37 °C and 200 strokes/min for 2 h. Intestinal digestion was simulated by correcting the pH of the samples with 5 M NaOH to 6.9 and adjusting the volume to 50 mL with the addition of 20 mM sodium phosphate buffer (pH 6.9, 10 mM NaCl). After the addition of 100 mg of pancreatin from a porcine pancreas, each sample was transferred into a dialysis tube (12–14 kD, Spectra/Por) with 5 glass marbles. The tubes were sealed and suspended in sealed containers with 600 mL of 20 mM sodium phosphate buffer (pH 6.9, 10 mM NaCl).

The containers were incubated for 5 h in a shaking water bath at 37 °C and 200 strokes/min to simulate the intestinal phase. The dialysate (1 mL) was collected after 15, 30, 45, 60, 90, 120, 150, 180, 240, and 300 min from the start of incubation. Complete starch hydrolysis was carried out by adding 30 µL of 0.5 N acetic acid and 20 µL of a solution of AMG from *Aspergillus niger* (300 U/mL AMG in water) to 0.5 mL of dialysate and incubating the samples at 60 °C for 2 h. At the end of the intestinal phase, the glucose concentration derived from starch hydrolysis was quantified using an automatic glucose analyzer (model 2900, Yellow Springs Instrument Company, Yellow Springs, OH, USA). The rate of digested starch from the samples (expressed as the percentage of digested starch) was calculated for each time point as follows: % digested starch = (glucose concentration × 0.9/Av starch) × 100. In vitro digestions were performed in triplicate for each product.

2.6. Slowly Digestible Starch and Rapidly Digestible Starch Determination

The percentages of slowly and rapidly digestible starch (SDS and RDS, respectively) were analyzed according to the method proposed by Englyst and colleagues [31], with slight modifications. Briefly, samples were cooked and extruded through 7 mm holes of a hand-operated mincer (Sirius, Karl Krüger), and then 2 g of product underwent several enzymatic attacks. After adding 10 mL of pepsin–guar solution (5 g/L pepsin and 5 g/L guar in 0.05 M HCl), the samples were vortex-mixed and incubated in a shaking water bath (SW23, Julabo) at 37 °C and 180 strokes/min for 30 min. Ten milliliters of preheated (37 °C) 0.25 M sodium acetate were added to 5 glass marbles, and the tubes were mixed and placed in a water bath for 3 min to equilibrate the temperature. The enzyme mixture was prepared by dissolving 3.3 g of pancreatin in 22 mL of distilled water, and the tubes were centrifuged (3200 rpm for 10 min). The supernatant (15 mL) was collected; then, 3.6 mL of AMG and 37.5 mg of invertase from baker's yeast (*Saccharomyces cerevisiae*) were diluted in 3.06 mL of distilled water and added to the supernatant. Five milliliters of the enzyme mixture were added to each sample, and the samples were incubated in a water bath at 37 °C and 200 strokes/min. After 20 min and 120 min, 1 mL of hydrolysate was centrifuged (14,000 rpm for 5 min), and the supernatant was diluted in distilled water (diluted 1:10), then used to determine the total glucose concentration (G_{20} and G_{120}, respectively).

To determine the free sugar glucose (FSG), the method of Englyst and colleagues [31] was used with some modifications. Briefly, 2 g of each product was extruded through 8 mm holes of a hand-operated mincer (Sirius, Karl Krüger) and weighed into plastic flasks. After the addition of 25 mL of 0.1 M sodium acetate buffer (pH 5.2) and 5 glass marbles, samples were vortex-mixed and incubated in a water bath (SW23, Julabo) at 100 °C for 30 min. Samples were vortex-mixed again and cooled to 37 °C; then, 0.153 mL of invertase solution (12.3 mg/mL invertase from baker's yeast (S. *cerevisiae*) in water) was added to the samples before incubation at 37 °C and 200 strokes/min for 30 min. One milliliter of hydrolysate was collected and centrifuged (14,000 rpm for 5 min), and the supernatant was used to determine the FSG.

The levels of rapidly available glucose (RAG), slowly available glucose (SAG), RDS, SDS, and Av starch were calculated as described by Englyst and colleagues [34].

The glucose amounts released after 20 and 120 min and after FSG analysis were quantified by means of an automatic glucose analyzer (model 2900, Yellow Springs Instrument Company, Yellow Springs, OH, USA). The analyses were performed in quadruplicate for each sample.

2.7. Statistical Analysis

All the results are expressed as the mean ± standard deviation (SD). The data distribution was assessed by means of the Shapiro–Wilk test, and the differences among results were studied by analysis of variance through one-way ANOVA and Bonferroni post hoc testing. Statistical significance was determined at $p < 0.05$, and the analyses were performed using SPSS Statistics software (version 26, IBM, Armonk, NY, USA).

3. Results

3.1. Starch Digestibility of Pasta Samples

The percentages of digested starch during the 5 h intestinal phase digestion are shown in Figure 1, with a focus on the percentages of starch digested after 120 and 300 min (Figure 1). After 120 min of simulated digestion, the BF pasta presented the highest percentage of starch digestion (39.67 ± 1.54%), followed by the DK (38.77 ± 3.63%), while the CTRL pasta showed the lowest percentage (38.66 ± 2.32%). After 300 min of digestion, the different starch digestibility percentages were 81.24 ± 1.62%, 82.44 ± 2.09%, and 78.42 ± 2.02% for BF, DK, and CTRL, respectively. Although a trend of reduced starch release during the digestion of CTRL compared to the two pasta samples produced via a non-conventional process was observed, no statistically significant differences among the samples were evident ($p > 0.05$).

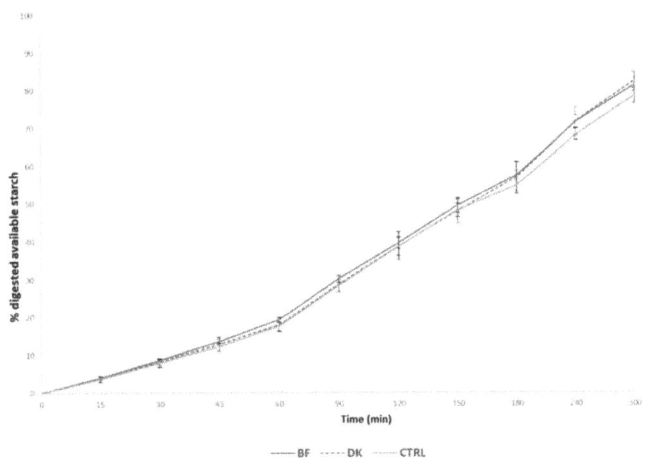

Figure 1. Lines represent the digested starch reported as a percentage of the total available starch during the in vitro digestion of each test food. Values are reported as the mean ± standard deviation (SD) (n = 3). BF: pasta produced by enriching semolina with a durum wheat bran fraction; CTRL: control, pasta produced by traditional milling; DK: pasta produced by using micronized debranned kernels. Statistical analysis was performed via one-way ANOVA and Bonferroni post hoc testing ($p < 0.05$).

3.2. Slowly and Rapidly Digestible Starch Determination

The values of RAG, SAG, RDS, SDS, Av starch, and FSG are reported in Table 1. BF had the highest amounts of RAG (17.81 ± 0.91 g/100 g) and RDS (15.71 ± 0.99 g/100 g) compared to the DK and CTRL samples. On the contrary, the CTRL pasta showed the highest values for SAG, SDS, and Av starch (8.94 ± 1.23, 8.04 ± 1.11, and 20.16 ± 0.98 g/100 g, respectively).

Table 1. The RAG, SAG, SDS, RDS, Av starch, and FSG values of the pasta samples assessed by an in vitro method. Values are expressed as the mean ± SD (n = 4).

Pasta Products	RAG (g/100 g)	SAG (g/100 g)	SDS (g/100 g)	RDS (g/100 g)	Av Starch (g/100 g)	FSG (g/100 g)
BF	17.81 ± 0.91 [a]	1.95 ± 0.72 [c]	1.75 ± 0.65 [b]	15.71 ± 0.99 [a]	17.46 ± 0.37 [b]	0.36 ± 0.01 [b]
DK	14.88 ± 0.32 [b]	4.78 ± 0.39 [b]	4.30 ± 0.35 [b]	12.91 ± 0.28 [b]	17.21 ± 0.21 [b]	0.54 ± 0.01 [a]
CTRL	13.77 ± 0.27 [b]	8.94 ± 1.23 [a]	8.04 ± 1.11 [a]	12.11 ± 0.24 [b]	20.16 ± 0.98 [a]	0.31 ± 0.00 [b]

Av starch: available starch; BF: pasta produced by enriching semolina with a durum wheat bran fraction; CTRL: control, pasta produced by traditional milling; DK: pasta produced by using micronized debranned kernels; FSG: free sugar glucose; RAG: rapidly available glucose; RDS: rapidly digestible starch; SAG: slowly available glucose; SDS: slowly digestible starch. Data in the same column with different letters indicate significant differences at $p < 0.05$, according to Bonferroni post hoc testing.

The percentage contributions of RDS and SDS to the Av starch (considered as 100%) are graphically reported in Figure 2. The ratio between SDS and Av starch was different among the three samples; in fact, the control (CTRL) showed a percentage four-fold higher than that for the BF and 1.5-fold higher than that for the DK sample.

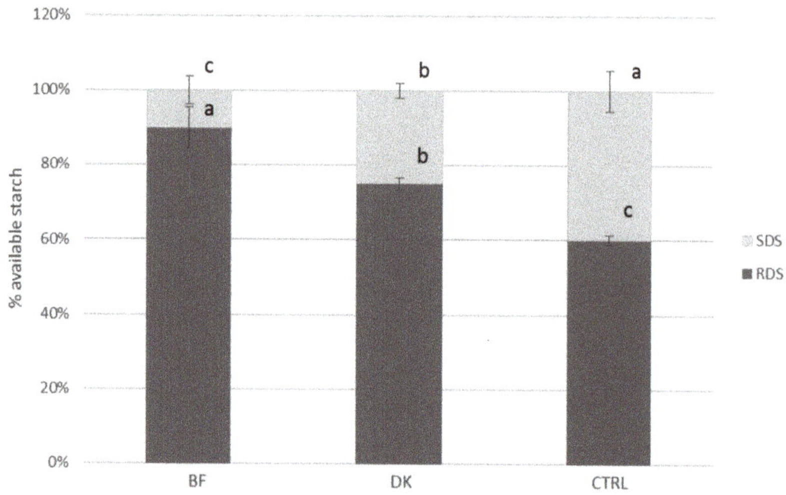

Figure 2. RDS and SDS expressed as percentages of the available starch (mean ± SD) (n = 4) for the pasta samples. BF: pasta produced by enriching semolina with a durum wheat bran fraction; CTRL: control, pasta produced by traditional milling; DK: pasta produced by using micronized debranned kernels; RDS: rapidly digestible starch; SDS: slowly digestible starch. Statistical analysis was performed via one-way ANOVA and Bonferroni post hoc testing. Different letters indicate statistical significance ($p < 0.05$).

4. Discussion

The present study aimed at evaluating the in vitro digestibility of three different pasta samples made from the cultivar Normanno using different types of pasta-making processes. In particular, this study explored the impact of the use of debranned products on the starch digestibility of pasta by applying two different methods for investigating the digestibility of starch, both recognized as suitable for assessing starch digestibility in food products [35]. The results of the present work reveal that the CTRL pasta showed the highest value of SDS/Av starch and the lowest value of RDS/Av starch, which could reflect a lower glycemic response in vivo compared to the BF and DK samples [36]. The process applied for enriching the pasta with fiber and bioactives seems ineffective in maintaining a compact starch matrix. This could be ascribable to the different process applied for the production of pasta samples. This may lead to the presence of a high amount of dietary fiber in BF pasta that could affect the structure of the food matrix, interfering with the formation of the gluten network, water absorption, and starch granule accessibility [37]. The DK pasta showed intermediate values of SDS/Av starch and RDS/Av starch, which could be due to the smaller size of bran particles. Moreover, it was previously shown that micronized samples of barley and maize pasta exhibited increased starch digestibility, and this could be attributed to starch gelatinization during micronization without significant retrogradation during storage [38,39]. When the method proposed by Brighenti and colleagues was applied, the same trend of starch digestibility was observed, with the CTRL showing a slightly lower starch release during digestion, even though no statistically significant differences were observed. However, the differences obtained between the two in vitro methods could be ascribable to the difference in the oral phase and in the incubation system [40]. Starch bioaccessibility is indeed influenced by, among other things, the mastication process. Through mastication, food is broken into smaller pieces, and the rate of digestion also depends on the time and the intensity of chewing, mainly due to the contact of the food surface with saliva α-amylase, responsible for the first starch hydrolysis [41]. However, the structural properties of pasta products can lead to different breakdown patterns during mastication and, consequently, different in vitro digestibility [42].

The debranning process improves the yield and degree of semolina refinement and enhances the nutritional value of the end-products [21], allowing us to obtain BFs with high fiber and bioactive contents. This process was used in combination with micronization, a technological process which enables a reduction in the food matrix into a fine powder, improving the bioaccessibility of the bioactive compounds [43] and making the bran particles smaller, lowering the impact of the dietary fiber on the gluten matrix. However, several studies have demonstrated the impact of fiber on the rate of starch digestion, with non-starch polysaccharides being responsible for discontinuity in the network, leading to faster hydrolysis. The addition of fiber to durum wheat pasta can interfere with the gluten structure, thus disrupting the continuity of the protein–starch matrix and making the starch granules more susceptible to enzymatic degradation [37]. In addition to fiber, starch digestibility in cereal-based products can be affected by several other factors, including the type and source of starch, the presence of protein matrixes, and the processing method [44].

In a previous study [24], pasta produced with debranning products (DK and BF) presented higher contents of phenolic compounds and other bioactives compared to traditional pasta, with minimal effects on its sensory properties [25]. However, the effect on starch digestibility related to the presence of debranning products has not been investigated. Bioactives present in foods may also play a key role in reducing the post-prandial glycemic response of carbohydrate-rich foods in vivo [45]. This can be mediated by the direct inhibition of starch enzymatic digestion, but also by other physiological mechanisms, such as inhibition of the absorption and potentially increased insulin release at the β-pancreatic level [45]. Therefore, we can hypothesize that a higher amount of bioactives in BF and DK samples may lead to a reduced glycemic response in vivo, which is strictly dependent on the bioaccessibility of phenolic compounds.

Diets exerting a low glycemic response favorably affect glucose metabolism and health status [46] and have been associated with a lower risk of many chronic diseases, such as type 2 diabetes and other cardiometabolic diseases, compared to high-GI diets [18,46,47]. In this scenario, pasta is a milestone in several healthy eating patterns, such as the Mediterranean diet, and its consumption is associated with several health benefits [17–19,48]. The consumption of pasta has decreased in recent years, probably because its consumption is wrongly associated with the myth of weight gain from the consumption of carbohydrate-rich foods [49]. However, recent studies have investigated the effect of pasta consumption on body weight and disease risk. Several publications have emphasized the beneficial role of pasta consumption on the obesity epidemic and cardiometabolic risk factors both in healthy subjects and in obese and diabetic patients [20–22,37]. In particular, pasta consumption, in the context of a low-GI diet, has shown to be involved in the reduction in body weight and markers of adiposity, such as waist circumference and waist-to-hip ratio [19]. Similar results emerged from another clinical trial recently conducted on obese patients [17]. In this study, the consumption of a hypocaloric diet characterized by an intake of at least five portions per week of pasta led to a higher reduction in body weight compared to the consumption of a lower amount of pasta ($\leq$3 portions/week). Finally, patients with type 2 diabetes did not show worse glucose control, measures of adiposity, or major cardiovascular risk factors when pasta was included in their diet within the recommended consumption amount [48]. Based on this evidence, the present work focused on the importance of producing high-quality pasta, taking into account several nutritional factors.

Finally, it is worth noting that "pasta" as a category includes a large number of heterogeneous types of products which differ in shape, ingredients, and nutritional composition, eliciting different responses in humans [50]. Considering the crucial role of pasta in several dietary patterns and the overall high worldwide consumption, it is worth investigating strategies for maximizing the nutritional quality of this product, preserving its naturally high SDS content. Therefore, the investigation of starch digestion in non-conventional pasta-making processes seems to be a potential strategy to obtain pasta with high contents of SDS, fiber, micronutrients, and bioactives to increase its health-related beneficial properties.

5. Conclusions

The starch digestibility of the BF and DK pasta samples after in vitro simulated gastrointestinal digestion was affected by the non-conventional pasta-making process applied to the samples, as compared to the traditional process. Enriching pasta with bran fractions seems to affect starch digestibility by decreasing the SDS content, while this effect was less obvious in pasta made with micronized debranned kernels. However, the traditional pasta-making process led to the product with the highest amount of SDS. This result may depend on the different structural properties of the three pasta samples, leading to differences in the rate of starch digestion in vitro. Therefore, further investigations should focus on the evaluation of the microstructure of pasta samples, and in vivo studies are needed to clarify the role of debranning products and the micronization process on the GI of pasta products. Moreover, future studies should also investigate the bioaccessibility and bioavailability of polyphenolic compounds from BF and DK pasta samples, to obtain a clearer picture of the overall nutritional quality of these products. Considering that dietary guidelines suggest the consumption of low-GI and high-fiber foods, it is strongly advisable to further explore new technologies for preserving the low impact of pasta on post-prandial glycaemia but guaranteeing the presence of fiber, the global consumption of which is still lower than the recommendations.

Author Contributions: Conceptualization, D.M. and M.D.; methodology, M.D. and D.M.; formal analysis, L.B., M.C., and F.T.; investigation, B.B. and M.C.; resources, F.S. and M.G.D.; data curation, R.D. and L.B.; writing—original draft preparation, R.D., M.D., D.M.; writing—review and editing, B.B., L.B., F.S., F.T., and M.G.D.; supervision, D.M. All authors have read and agreed to the published version of the manuscript.

Funding: This research was funded by the AGER Project, grant no. 2010-0262.

Institutional Review Board Statement: Not applicable.

Informed Consent Statement: Not applicable.

Conflicts of Interest: The authors declare no conflict of interest.

References

1. International Pasta Organisation (IPO). *The World Pasta Industry Status Report*; International Pasta Organization: Rome, Italy, 2019; Available online: http://www.internationalpasta.org (accessed on 1 March 2021).
2. Consiglio per la ricerca in agricoltura e l'analisi dell'economia agraria (CREA). Guidelines for Healthy Italian Nutrition. 2019. Available online: https://www.crea.gov.it/web/alimenti-e-nutrizione/-/linee-guida-per-una-sana-alimentazione-2018 (accessed on 1 March 2021).
3. Leclercq, C.; Arcella, D.; Piccinelli, R.; Sette, S.; Le Donne, C. The Italian national food consumption survey INRAN-SCAI 2005–06: Main results in terms of food consumption. *Public Health Nutr.* **2009**, *12*, 2504–2532. [CrossRef] [PubMed]
4. Società Italiana di Nutrizione Umana. *Larn—Livelli di assunzione di Riferimento di Nutrienti ed Energia per la Popolazione Italiana*, 4th ed.; Società Italiana di Comunicazione Scientifica e Sanitaria (SICS): Milano, Italy, 2014.
5. Saura-Calixto, F.; Goñi, I. Definition of the Mediterranean diet based on bioactive compounds. *Crit. Rev. Food Sci. Nutr.* **2009**, *49*, 145–152. [CrossRef] [PubMed]
6. Augustin, L.S.A.; Aas, A.M.; Astrup, A.; Atkinson, F.S.; Baer-Sinnott, S.; Barclay, A.W.; Brand-Miller, J.C.; Brighenti, F.; Bullo, M.; Buyken, A.E.; et al. Dietary fibre consensus from the International Carbohydrate Quality Consortium (ICQC). *Nutrients* **2020**, *12*, 2553. [CrossRef] [PubMed]
7. Ciudad-Mulero, M.; Fernández-Ruiz, V.; Matallana-González, M.C.; Morales, P. Dietary fiber sources and human benefits: The case study of cereal and pseudocereals. *Adv. Food Nutr. Res.* **2019**, *90*, 83–134.
8. Del Bo, C.; Bernardi, S.; Marino, M.; Porrini, M.; Tucci, M.; Guglielmetti, S.; Cherubini, A.; Carrieri, B.; Kirkup, B.; Kroon, P.; et al. Systematic review on polyphenol intake and health outcomes: Is there sufficient evidence to define a health-promoting polyphenol-rich dietary pattern? *Nutrients* **2019**, *11*, 1355.
9. Sette, S.; D'Addezio, L.; Piccinelli, R.; Hopkins, S.; Le Donne, C.; Ferrari, M.; Mistura, L.; Turrini, A. Intakes of whole grain in an Italian sample of children, adolescents and adults. *Eur. J. Nutr.* **2017**, *56*, 521–533. [CrossRef]
10. Whole Grain Statistics. Available online: https://wholegrainscouncil.org/newsroom/whole-grain-statistics (accessed on 1 March 2021).
11. Neo, J.E.; Brownlee, I.A. Wholegrain food acceptance in young Singaporean adults. *Nutrients* **2017**, *9*, 371. [CrossRef]
12. Van der Kamp, J.W.; Poutanen, K.; Seal, C.J.; Richardson, D.P. The healthgrain definition of 'whole grain'. *Food Nutr. Res.* **2014**, *58*. [CrossRef]
13. Foster, S.; Beck, E.; Hughes, J.; Grafenauer, S. Whole grains and consumer understanding: Investigating consumers' identification, knowledge and attitudes to whole grains. *Nutrients* **2020**, *12*, 2170. [CrossRef]
14. Barrett, E.M.; Foster, S.I.; Beck, E.J. Whole grain and high-fibre grain foods: How do knowledge, perceptions and attitudes affect food choice? *Appetite* **2020**, *149*, 104630. [CrossRef]
15. Petitot, M.; Abecassis, J.; Micard, V. Structuring of pasta components during processing: Impact on starch and protein digestibility and allergenicity. *Trends Food Sci. Tech.* **2009**, *20*, 521–532. [CrossRef]
16. Scazzina, F.; Dall'Asta, M.; Casiraghi, M.C.; Sieri, S.; Del Rio, D.; Pellegrini, N.; Brighenti, F. Glycemic index and glycemic load of commercial Italian foods. *Nutr. Metab. Cardiovasc. Dis.* **2016**, *26*, 419–429. [CrossRef] [PubMed]
17. Rosi, A.; Tesan, M.; Cremonini, A.; Biasini, B.; Bicchieri, L.; Cossu, M.; Brighenti, F.; Dall'Aglio, E.; Scazzina, F. Body weight of individuals with obesity decreases after a 6-month high pasta or low pasta Mediterranean diet weight-loss intervention. *Nutr. Metab. Cardiovasc. Dis.* **2020**, *30*, 984–995. [CrossRef] [PubMed]
18. Huang, M.; Li, J.; Ha, M.A.; Riccardi, G.; Liu, S. A systematic review on the relations between pasta consumption and cardio-metabolic risk factors. *Nutr. Metab. Cardiovasc. Dis.* **2017**, *27*, 939–948. [CrossRef] [PubMed]
19. Chiavaroli, L.; Kendall, C.W.C.; Braunstein, C.R.; Blanco Mejia, S.; Leiter, L.A.; Jenkins, D.J.A.; Sievenpiper, J.L. Effect of pasta in the context of low-glycaemic index dietary patterns on body weight and markers of adiposity: A systematic review and meta-analysis of randomised controlled trials in adults. *BMJ Open* **2018**, *8*, e019438. [CrossRef] [PubMed]
20. Patel, S. Cereal bran fortified-functional foods for obesity and diabetes management: Triumphs, hurdles and possibilities. *J. Funct. Foods* **2015**, *14*, 255–269. [CrossRef]
21. Dexter, J.E.; Wood, P.J. Recent applications of debranning of wheat before milling. *Trends Food Sci. Tech.* **1996**, *7*, 35–41. [CrossRef]
22. Martini, D.; D'Egidio, M.G.; Nicoletti, I.; Corradini, D.; Taddei, F. Effects of durum wheat debranning on total antioxidant capacity and on content and profile of phenolic acids. *J. Funct. Foods* **2015**, *17*, 83–92. [CrossRef]
23. Mousia, Z.; Edherly, S.; Pandiella, S.S.; Webb, C. Effect of wheat pearling on flour quality. *Food Res. Int.* **2004**, *37*, 449–459. [CrossRef]
24. Ciccoritti, R.; Taddei, F.; Nicoletti, I.; Gazza, L.; Corradini, D.; D'Egidio, M.G.; Martini, D. Use of bran fractions and debranned kernels for the development of pasta with high nutritional and healthy potential. *Food Chem.* **2017**, *225*, 77–86. [CrossRef]
25. Martini, D.; Ciccoritti, R.; Nicoletti, I.; Nocente, F.; Corradini, D.; D'Egidio, M.G.; Taddei, F. From seed to cooked pasta: Influence of traditional and non-conventional transformation processes on total antioxidant capacity and phenolic acid content. *Int. J. Food Sci. Nutr.* **2018**, *69*, 24–32. [CrossRef]
26. Laca, A.; Mousia, Z.; Díaz, M.; Webb, C.; Pandiella, S.S. Distribution of microbial contamination within cereal grains. *Food Eng.* **2006**, *72*, 332–338. [CrossRef]
27. Tibola, C.S.; Guarienti, E.M.; Dias, A.R.G.; Nicolau, M.; Devos, R.J.B.; Teixeira, D.D. Effect of debranning process on deoxynivalenol content in whole-wheat flours. *Cereal Chem.* **2019**, *96*, 717–724. [CrossRef]

28. Giordano, D.; Blandino, M. Arsenic, lead and cadmium distribution in the pearled fractions of different winter wheat cultivars (*Triticum aestivum* L.). *J. Cereal Sci.* **2018**, *80*, 94–101. [CrossRef]
29. Foster-Powell, K.; Holt, S.H.A.; Brand-Miller, J.C. International table of glycemic index and glycemic load values: 2002. *Am. J Clin. Nutr.* **2002**, *76*, 5–56. [CrossRef] [PubMed]
30. Atkinson, F.S.; Foster-Powell, K.; Brand-Miller, J.C. International tables of glycemic index and glycemic load values: 2008. *Diabetes Care* **2008**, *31*, 2281–2283. [CrossRef] [PubMed]
31. Englyst, K.N.; Englyst, H.N.; Hudson, G.J.; Cole, T.J.; Cummings, J.H. Rapidly available glucose in foods: An in vitro measurement that reflects the glycemic response. *Am. J. Clin. Nutr.* **1999**, *69*, 448–454. [CrossRef] [PubMed]
32. Brighenti, F.; Pellegrini, N.; Casiraghi, M.C.; Testolin, G. In vitro studies to predict physiological effects of dietary fibre. *Eur. J. Clin. Nutr.* **1995**, *49*, S81–S88.
33. Zaupa, M.; Scazzina, F.; Dall'Asta, M.; Calani, L.; Del Rio, D.; Bianchi, M.A.; Melegari, C.; De Albertis, P.; Tribuzio, G.; Pellegrini, N.; et al. In vitro bioaccessibility of phenolics and vitamins from durum wheat aleurone fractions. *J. Agric. Food Chem.* **2014**, *62*, 1543–1549. [CrossRef]
34. Englyst, K.N.; Hudson, G.J.; Englyst, H.N. Starch analysis in food. In *Encyclopedia of Analytical Chemistry*; Meyers, R.A., Ed.; John Wiley & Sons Ltd.: Chichester, UK, 2000; pp. 4246–4262.
35. EFSA Panel on Dietetic Products, Nutrition and Allergies (NDA). Scientific Opinion on the substantiation of a health claim related to "slowly digestible starch in starch-containing foods" and "reduction of post-prandial glycaemic responses" pursuant to Article 13(5) of Regulation (EC) No 1924/2006. *EFSA J.* **2011**, *9*, 2292.
36. Englyst, H.N.; Veenstra, J.; Hudson, G.J. Measurement of rapidly available glucose (RAG) in plant foods: A potential in vitro predictor of the glycaemic response. *Br. J. Nutr.* **1996**, *75*, 327–337. [CrossRef]
37. Gallo, V.; Romano, A.; Masi, P. Does the presence of fibres affect the microstructure and in vitro starch digestibility of commercial Italian pasta? *Food Struct.* **2020**, *24*, 100139. [CrossRef]
38. Emami, S.; Meda, V.; Pickard, M.D.; Tyler, R.T. Impact of micronization on rapidly digestible, slowly digestible, and resistant starch concentrations in normal, high-amylose, and waxy barley. *J. Agric. Food Chem.* **2010**, *58*, 9793–9799. [CrossRef] [PubMed]
39. Deepa, C.; Sarabhai, S.; Prabhasankar, P.; Hebbar, H.U. Effect of micronization of maize on quality characteristics of pasta. *Cereal Chem.* **2017**, *94*, 840–846. [CrossRef]
40. Germaine, K.A.; Samman, S.; Fryirs, C.G.; Griffiths, P.J.; Johnson, S.K.; Quail, K.J. Comparison of in vitro starch digestibility methods for predicting the glycaemic index of grain foods. *J. Sci. Food Agric.* **2008**, *88*, 652–658. [CrossRef]
41. Blanquet-Diot, S.; François, O.; Denis, S.; Hennequin, M.; Peyron, M.A. Importance of oral phase in in vitro starch digestibility related to wholegrain versus refined pastas and mastication impairment. *Food Hydrocoll.* **2021**, *112*, 106277. [CrossRef]
42. Alam, S.A.; Pentikäinen, S.; Närväinen, J.; Katina, K.; Poutanen, K.; Sozer, N. The effect of structure and texture on the breakdown pattern during mastication and impacts on in vitro starch digestibility of high fibre rye extrudates. *Food Funct.* **2019**, *10*, 1958–1973. [CrossRef] [PubMed]
43. Hemery, Y.M.; Anson, N.M.; Havenaar, R.; Haenen, G.R.M.M.; Noort, M.W.J.; Rouau, X. Dry-fractionation of wheat bran increases the bioaccessibility of phenolic acids in breads made from processed bran fractions. *Food Res. Int.* **2010**, *43*, 1429–1438. [CrossRef]
44. Singh, J.; Dartois, A.; Kaur, L. Starch digestibility in food matrix: A review. *Trends Food Sci. Technol.* **2010**, *21*, 168–180. [CrossRef]
45. Dall'Asta, M.; Del Rio, D.; Tappy, L.; Potì, F.; Agostoni, C.; Brighenti, F. Critical and emerging topics in dietary carbohydrates and health. *Int. J. Food Sci. Nutr.* **2020**, *71*, 286–295. [CrossRef]
46. Augustin, L.S.; Kendall, C.W.; Jenkins, D.J.; Willett, W.C.; Astrup, A.; Barclay, A.W.; Bjorck, I.; Brand-Miller, J.C.; Brighenti, F.; Buyken, A.E.; et al. Glycemic index, glycemic load and glycemic response: An international scientific consensus summit from the International Carbohydrate Quality Consortium (ICQC). *Nutr. Metab. Cardiovasc. Dis.* **2015**, *25*, 795–815. [CrossRef]
47. Giacco, R.; Costabile, G.; Della Pepa, G.; Anniballi, G.; Griffo, E.; Mangione, A.; Cipriano, P.; Viscovo, D.; Clemente, G.; Landberg, R.; et al. A whole-grain cereal-based diet lowers postprandial plasma insulin and triglyceride levels in individuals with metabolic syndrome. *Nutr. Metab. Cardiovasc. Dis.* **2014**, *24*, 837–844. [CrossRef]
48. Vitale, M.; Masulli, M.; Rivellese, A.A.; Bonora, E.; Babini, A.C.; Sartore, G.; Corsi, L.; Buzzetti, R.; Citro, G.; Baldassarre, M.P.A.; et al. Pasta consumption and connected dietary habits: Associations with glucose control, adiposity measures, and cardiovascular risk factors in people with type 2 diabetes-TOSCA IT study. *Nutrients* **2019**, *12*, 101. [CrossRef] [PubMed]
49. Montel Press Office. Italy's Love of Pasta Goes off the Boil: Sales of Pasta Fall by 2%. Available online: https://www.mintel.com/press-centre/food-and-drink/italys-love-of-pasta-goes-off-the-boil (accessed on 1 March 2021).
50. Dello Russo, M.; Spagnuolo, C.; Moccia, S.; Angelino, D.; Pellegrini, N.; Martini, D.; On Behalf of the Italian Society of Human Nutrition Sinu Young Working Group. Nutritional quality of pasta sold on the Italian market: The food labelling of Italian products (flip) study. *Nutrients* **2021**, *13*, 171. [CrossRef] [PubMed]

foods

MDPI

Article

Innovative Milling Processes to Improve the Technological and Nutritional Quality of Parboiled Brown Rice Pasta from Contrasting Amylose Content Cultivars

Federica Taddei, Elena Galassi, Francesca Nocente and Laura Gazza *

CREA Research Centre for Engineering and Agro-Food Processing, Via Manziana 30, 00189 Rome, Italy; federica.taddei@crea.gov.it (F.T.); elena.galassi@crea.gov.it (E.G.); francesca.nocente@crea.gov.it (F.N.)
* Correspondence: laura.gazza@crea.gov.it

Abstract: The demand for gluten-free products, including pasta, is increasing and rice pasta accounts for the largest share of this market. Usually, the production of rice pasta requires additives or specific technological processes able to improve its texture, cooking quality, and sensory properties. In this work, two rice cultivars, with different amylose content, were subjected to parboiling, micronization, and flour air fractionation to obtain brown rice pasta, without any supplement but rice itself. In particular, two types of pasta (spaghetti shape) were produced, one from 100% micronized wholemeal, and the other from refined rice flour replaced with 15% of the air-fractionated fine fraction. Regardless of the cultivar, pasta from wholemeal micronized flour showed higher protein and fiber content than refined flour enriched with fine fraction, whereas no differences were revealed in resistant starch and antioxidant capacity. Pasta from the high amylose content genotype showed the highest resistant starch content and the lowest predicted glycemic index along with sensorial characteristics as good as durum semolina pasta in fine fraction enriched pasta. Besides the technological processes, pasta quality was affected the most by the genotype, since pasta obtained from high amylose cv Gladio resulted in the best in terms of technological and sensory quality.

Keywords: brown rice; micronization; air fractionation; parboiling; rice pasta

Citation: Taddei, F.; Galassi, E.; Nocente, F.; Gazza, L. Innovative Milling Processes to Improve the Technological and Nutritional Quality of Parboiled Brown Rice Pasta from Contrasting Amylose Content Cultivars. *Foods* **2021**, *10*, 1316. https://doi.org/10.3390/foods10061316

Academic Editor: Robert G. Gilbert, Alessandra Marti and Zhengyu Jin

Received: 8 April 2021
Accepted: 4 June 2021
Published: 8 June 2021

1. Introduction

Rice (*Oryza sativa* L.) is the second most important staple food crop, after wheat, and currently sustains half of the world's population [1]. Indeed, it contains carbohydrates (75–80%), proteins (7–8%), lipids (3%) and is also rich in dietary fiber, minerals, and vitamins, especially when consumed as wholegrain [2].

Owing to the real or presumed increase in gluten intolerances along with changing consumer preferences for more digestible foods, the demand for gluten-free products, including pasta, is increasing [3,4]. Among the gluten-free pasta, currently, rice pasta accounts for a higher value of the gluten-free pasta market share [4], due to its bland flavor, high digestibility, and hypoallergenic properties [5]. Usually, the production of rice pasta requires additives, such as proteins, gums, and emulsifiers, or specific technological process, such as extrusion cooking and hydrothermal treatments, which modify the functional properties of starch and protein, improving texture, cooking quality, and sensory properties of the cooked pasta [6,7].

Rice starch characteristics influence the processability and the technological properties of rice the most [8], mainly the ratio of amylose and amylopectin constituents. Understanding the characteristics of rice starch is very important for optimizing industrial end-products and providing consumers with suitable rice cultivars with enhanced health benefits. Brown rice noodles from high amylose content genotypes exhibited better texture and cooking quality [9]; moreover, starch granules rich in amylose resulted in a more crystalline structure than those with low amylose content. Consequently, they do not swell or gelatinize as readily upon cooking and, therefore, are digested more slowly, resulting in lower blood glucose and insulin responses than low-amylose content rice varieties. For this reason, the intake of high-amylose rice foods has been considered more desirable for individuals with impaired glucose metabolism [10].

Amongst the technological processes usually applied to rice, parboiling is able to modify the physicochemical properties of starch, avoiding the use of additives, such as texturing proteins, gums, and emulsifiers [11]. Indeed, it is reported that parboiling, alters the structural properties of rice starch, leading from crystalline to amorphous form and resulting in the highly compact and translucent endosperm and improving the sensory and cooking qualities, as well as the texture of the rice noodles [12]. Parboiling induces lipid-amylose complexes synthesis, aggregation of soluble proteins, resulting in a reduction of starch swelling and amylose leaching during cooking, in a decrease in stickiness, and in an increase in hardness [13,14]. Upon parboiling, an improvement of nutritional properties of rice also occurs, due to the migration of vitamins and minerals towards the endosperm, together with an increase in the levels of resistant starch (RS) [15,16], that appears to confer considerable benefits to human colonic health [17,18]. The use of flour from parboiled rice as raw material for pasta products [19], allowed to obtain pasta with a good cooking behavior due to the starch arrangements in the product [20].

Despite the higher nutritional value, brown rice is consumed less than white rice because of low consumer acceptability and of problematic technological aspects [21]. Mild separation technologies such as air flow fractionation could be applied for the production of wheat flour fractions, enriched in interesting healthy compounds, such as arabinoxilans, alkylresorcinols, and dietary fiber [22], to be added as ingredients for obtaining cereal-based functional foods, overcoming the technological and sensorial drawbacks of wholegrain. The aim of this work was to evaluate the behavior of two japonica rice cultivars, Gladio and Ronaldo, with contrasting amylose content, subjected to non-conventional transformation processes such as parboiling, micronization, and flour air fractionation to obtain rice pasta with increased nutritional and healthy potential. The effect of both the genotype and the technological process on the nutritional properties of raw materials and pasta samples, cooking quality, and starch hydrolysis index were investigated.

2. Materials and Methods

2.1. Plant Material

Japonica brown rice kernels (BR) of high amylose cv Gladio and intermediate-low amylose cv Ronaldo, classified according to Juliano [23], were kindly supplied by CREA-Research Centre for Cereal and Industrial Crops (Vercelli, Italy).

2.2. Technological Process

The flow chart of the processes applied is represented in Figure 1.

2.2.1. Parboiling Process

Brown rice kernels (BR) of cvs Gladio and Ronaldo were subjected to parboiling process following the method described by Hidalgo et al. [24]. In detail, kernels were conditioned for 4 h, until a moisture content of 15–16% was reached and then heated by steaming at 120 ± 1 °C, 2.1 bar, 10 min. The steamed kernels were dried for almost 48 h to reach 11% moisture, in an oven at 30 °C.

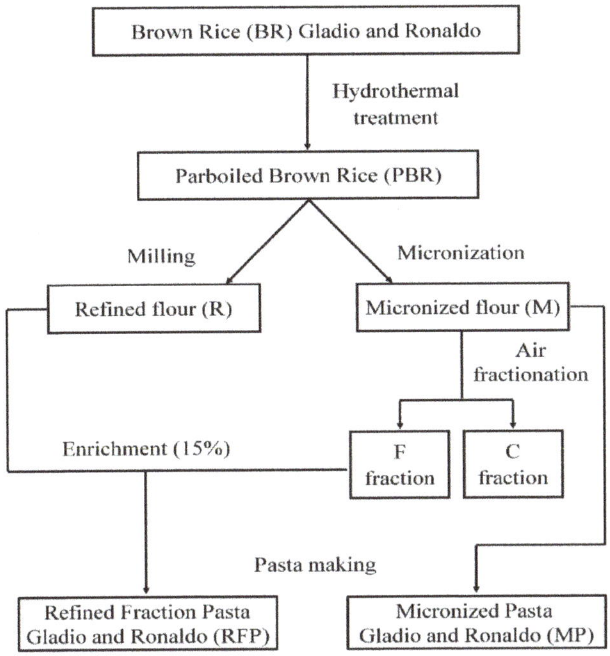

Figure 1. Flow chart of the technological processes applied to the two rice cultivars Gladio and Ronaldo.

2.2.2. Milling and Ultra-Fine Milling

Parboiled brown rice (PBR) was ground by a milling pilot plant (4RB BONA, Monza, Italy) to obtain refined flour (R). In addition, an aliquot of both parboiled (PBR) and not parboiled brown rice (BR) kernels were ground by the Cyclotec Laboratory Mill (FOSS, Hillerod, Denmark) 1 mm sieving and considered as the reference material. Milling processes were repeated twice. Ultra-fine milling (micronization) was applied on the parboiled kernels, in the KMX-500 device (Separ Microsystem, Brescia, Italy) at a frequency of 170 Hz to obtain micronized flours (M).

2.2.3. Air Fractionation

The micronized flours (M) were fractionated as described in Ciccoritti et al. [22] by a unit integrated turbo air separator (Separ Microsystem, Brescia, Italy) where an aspirating pump drives the air flow, which was modulated setting the inlet restriction valve at 250. The system sorted the flour in two fractions defined as coarse (C) and fine (F).

2.3. Pasta Making Process

Two pasta formulations were produced for each cultivar: (i) 100% micronized flour pasta (MP); (ii) 85% refined flour plus 15% F fraction pasta, (RFP) (Figure 1). Tap water was added to obtain a dough with 40% of moisture content. Pasta, spaghetti shape (1.6 mm diameter), was produced using an experimental press (NAMAD, Rome, Italy). Doughs were kneaded for 15 min at 50 °C. Rice pasta samples were dried horizontally by an experimental drier (AFREM, Lyon, France), applying a low temperature drying cycle for 18 h at 50 °C. Pasta from 100% semolina, from durum wheat cultivars Antalis and Svevo, was also produced following the same experimental conditions, except for the dough moisture content which was 34%.

2.4. Pasta Color, Cooking Quality and Sensory Test

The color of dried pasta samples was measured by Tristimulus Colorimeter, Chroma Meter CR-400 (Konica Minolta, Osaka, Japan), using the CIE-Lab color space coordinates L* (white-black), a* (red-green) and b* (yellow-blue), and the D65 illuminant (0° viewing angle geometry)

One hundred grams of rice pasta were added to 1L of boiling tap water without salt, according to the AACC method 66–50.01 [25] to obtain cooked rice pasta. Optimum cooking time (OCT) was evaluated according to D'Egidio et al. [26] and determined as when the white central core of the pasta just disappeared when squeezed between two glasses. Water absorption (WA), cooking loss (CL) and sensory analysis by a panel of three trained assessors, were evaluated as reported by Nocente et al. [27]. The sensorial judgment (SJ) was based on three textural characteristics: firmness, stickiness, and bulkiness. Each of the three parameters was evaluated by a score ranging from 10 to 100, by a trained and experienced panel of three assessors. The global value of the sensorial judgment (SJ) was the arithmetic mean of the three textural components [26].

2.5. Chemical Characterization

Chemical composition was assessed both on raw materials and dry pasta. Moisture was measured by a thermobalance (Sartorius MA 40, Goettingen, Germany) at 120 °C and all analytical data were expressed as dry weight (dw).

Total starch (TS) content was determined according to the Official Method 996.11 [28], by Total Starch Assay Kit (Megazyme, Bray, Ireland). Amylose content was determined using the Megazyme Amylose/Amylopectin assay kit. Resistant starch (RS) content was determined according to the Official Method 2002.02 [29], using Resistant Starch Assay Kit (Megazyme). Total dietary fiber (TDF) content was measured using the enzymatic kit Bioquant (Merck, Darmstadt, Germany) according to the Official Method 991.42 [30]. Ash content was determined according to the Official Method 08-01.01 [31]. Protein content was determined by micro-Kjeldhal nitrogen analysis, according to the ICC 105-2 method [32], using as conversion factor N $\times$ 6.25. Total antioxidant capacity (TAC) was determined according to Martini et al. [33].

Starch Hydrolysis Index and Predicted Glycemic Index

Starch hydrolysis was analysed following the method described by Goñi et al. [34]. One hundred milligram of cooked pasta samples were homogenized in HCl–KCl buffer pH 1.5 using an Ultra Turrax homogenizer (T25, Ika Labortechnik, Staufen, Germany). Then, samples were digested by pepsin from porcine gastrine mucosa (Merck) followed by α-amylase from porcine pancreas (Merck) and by amyloglucosidase from *Aspergillus niger* (Merck). Glucose concentration was measured using the glucose oxidase–peroxidase (GOPOD) kit (Megazyme). The rate of starch digestion was expressed as a percentage of the total starch hydrolyzed at different times [34]. To describe the kinetics of starch hydrolysis the area under the hydrolysis curve, hydrolysis index and expected glycemic index were estimated using the Goni [34] proposed equations.

2.6. Statistical Analysis

All analyses were performed in three replicates unless otherwise stated. Replicated results were expressed as mean ± standard deviation. Means were compared using the Kruskal–Wallis test to highlight significant differences ($p \leq 0.05$) among the different samples for each considered parameter, followed by Mann–Whitney test for paired comparison of the samples. Principal Component Analysis (PCA) was carried out to investigate the relationships among all pasta variables under study. Software PAST 4.02 (Oslo, Norway) was used to conduct data analysis.

3. Results and Discussion

3.1. Chemical Characterization of Raw Materials and Milling Products

In both cultivars, brown rice (BR), parboiled brown rice (PBR), and micronized flour (M) showed no significant differences in total starch (TS) content, being the mean value of about 80% (Figure 2A). Refined flour (R) resulted in the highest TS content (86.7 and 85.9% in cv Gladio and Ronaldo, respectively), as a consequence of the outmost layers' removal leading to an increased contribution of the amylaceous endosperm to the total weight [14]. Air fractionated C fraction presented TS values higher ($p \leq 0.05$) than F fraction in both cultivars (Figure 2A). This could be explained by the effect of the air fractionation process conditions that led to a major concentration of total starch in the milling fractions presenting flour with higher particle size and lower fiber content, as previously observed also in durum wheat [22].

The analysis of amylose content confirmed the high (cv Gladio) and intermediate-low (cv Ronaldo) amylose trait [35], showing a mean value of 30.8 and 24.6%, respectively (Figure 2B). Parboiling process applied to brown rice kernels determined a general increase in amylose, detectable to a major and significant ($p \leq 0.05$) extent in the intermediate-low amylose cv Ronaldo. This result could be ascribable to the rearrangements of amylose and amylopectin, upon the hydrothermal treatment, as already reported in [20]. In detail, a significant increase in comparison to BR and PBR was observed in refined parboiled flours (R) (34.7% and 30.6% in Gladio and Ronaldo, respectively), devoid of external layers, as above discussed for the TS content. In both cultivars, the C fraction showed a significantly higher amylose percentage than the F fraction (34.4% vs. 27.6% in cv Gladio and 27.0% vs. 22.2% in cv Ronaldo) (Figure 2B), as already observed for the TS. This result could be a consequence of the major structure breakage of starch granules occurring upon micronization mainly in the smallest mean size fraction (F), as also observed by Hossen et al. [36].

Resistant starch content was always significantly higher in cv Gladio than in cv Ronaldo, with a mean value of 0.879% in Gladio and of 0.365% in Ronaldo (Figure 2C). Though the method used [29] could implicate low accuracy (standard error higher than 5%) when applied to samples containing less than 2% resistant starch, it allowed discernment of the differences among all matrices analyzed and specifically between the two cultivars. Results are in accordance with previous findings in which amylose content was found to be positively correlated with resistant starch [35–37]. In detail, for parboiled brown rice, RS content increased in both cultivars, with a much higher magnitude in cv Gladio than in cv Ronaldo (+81% vs. +11%, respectively), with a significant increase ($p \leq 0.05$) only in cv Gladio. The observed increase is due to the effect of cooling after the parboiling process that led to starch retrogradation [38–40]. Sample R contained less RS than PBR in cv Gladio, whereas in cv Ronaldo, the tendency is the opposite (Figure 2C), likely due to the differences in amylose and the correlated RS content of the two genotypes, which could affect the response to the milling process. On the contrary, micronized flours (M) of both cultivars presented, as expected, RS values similar and not statistically different to those observed in PBR, suggesting that milling method had no effect on RS percentage. The comparison between F and C fractions revealed no significant differences in RS content in cv Ronaldo, but a significant decrease in F fraction was detected in cv Gladio (Figure 2C).

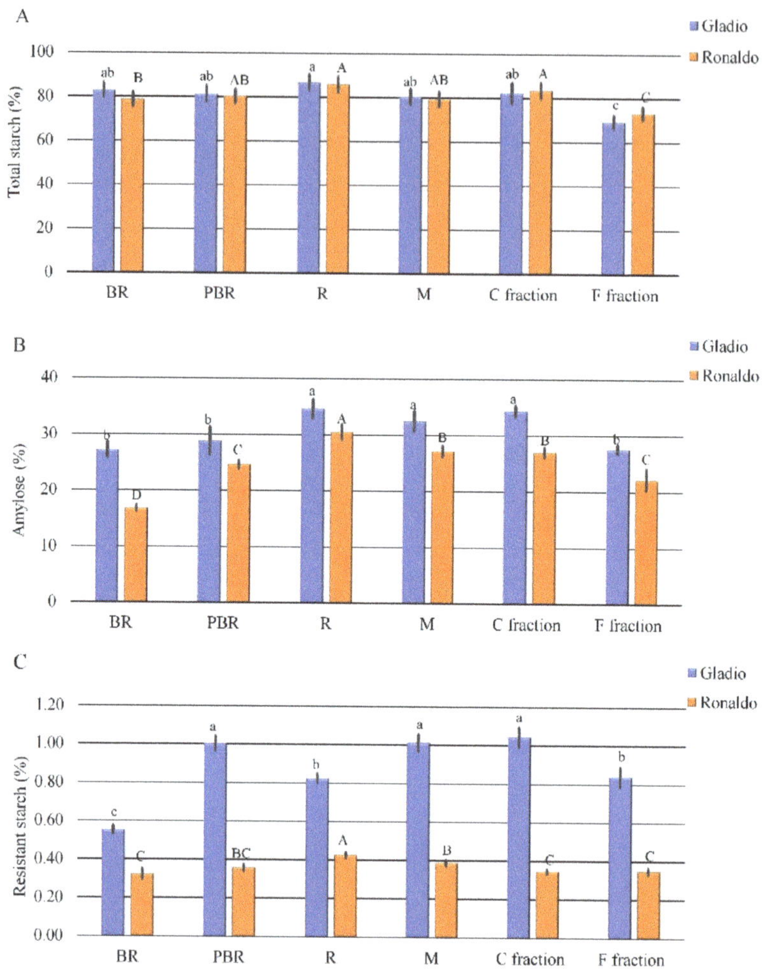

Figure 2. Total starch (**A**), amylose (**B**) and resistant starch (**C**) content in raw materials of rice cultivars Gladio and Ronaldo. BR = Brown Rice; PBR = Parboiled Brown Rice; R = Refined parboiled flour; M = Micronized parboiled flour; C = Coarse; F = Fine. Different letters indicate significant differences determined by the Mann–Whitney pairwise test ($p \leq 0.05$). Lower case letters refer to cv Gladio; upper case letters refer to cv Ronaldo. Results are expressed as ± standard deviation for three replications.

Protein content in brown rice kernels (BR) of both cvs was slightly higher than 8.0% (Figure 3A); after the parboiling process, a slight but not significant decrease in cv Gladio (−6%) and a significant increase in cv Ronaldo (+11.6%) was observed. Though, generally, proteins are reported to be less efficiently extracted from parboiling rice [41], due to leaching, breaking, and gelatinized starch entrapment that occurred over soaking and steaming, some authors [42,43] found a significant increase in the protein content in parboiled rice kernel. This opposite behavior is probably due to the different responses of the genotypes to the parboiling process. Because of the removal of the outer layers, where part of the proteins is located, in the refined flour (R), a not significant decrease in the protein content was observed in cv Gladio (−16.5%), whereas a significant decrease (−17.2%) was detected in cv Ronaldo (Figure 3A). Indeed, in the micronized wholemeal

(M), the protein content did not differ from that found in the parboiled kernels (PBR). The air fractionation process led to a protein content significant increase in F fraction with respect to M flour (+23% in Gladio and +8% in Ronaldo, Figure 3A).

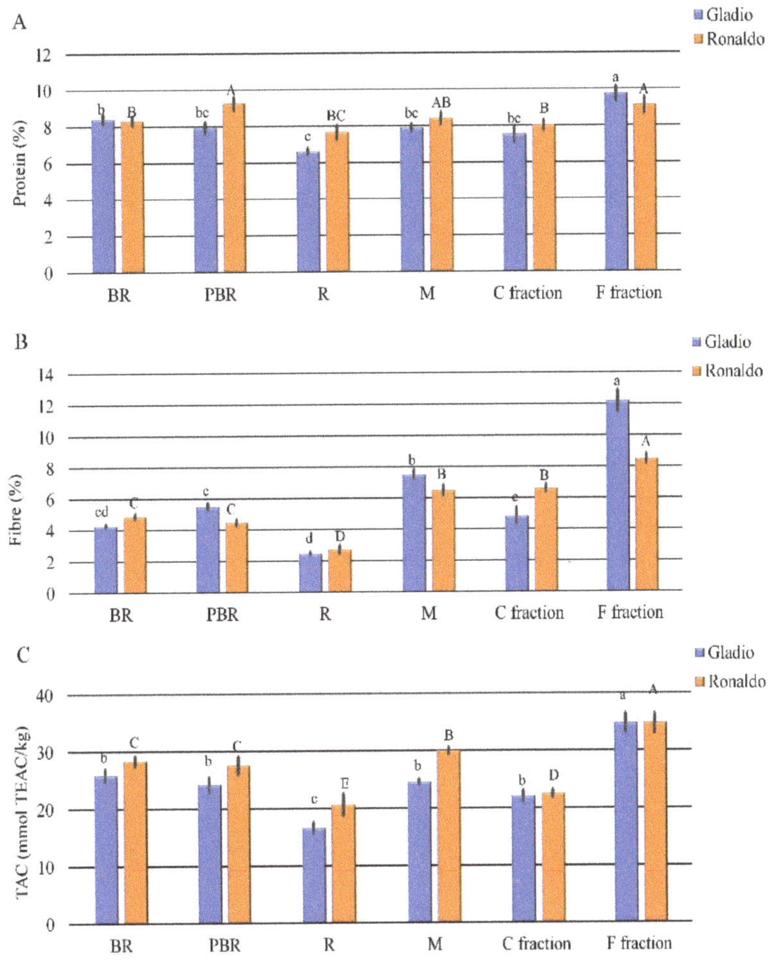

Figure 3. Protein (**A**) and fiber (**B**) content and Total antioxidant capacity (TAC) level (**C**) in raw materials of rice cultivars Gladio and Ronaldo. BR = Brown Rice; PBR = Parboiled Brown Rice; R = Refined parboiled flour; M = Micronized parboiled flour; C = Coarse; F = Fine; TEAC = trolox equivalent antioxidant capacity. Different letters indicate significant differences determined by the Mann–Whitney pairwise test ($p \leq 0.05$). Lower case letters refer to cv Gladio; upper case letters refer to cv Ronaldo. Results are expressed as ± standard deviation for three replications.

Upon parboiling process, only in cv Gladio a slight but not significant fiber content increase was determined in brown rice kernels (Figure 3B). As expected, refined flour (R) showed a significant decrease in fiber content in both cultivars (−55% and −39% in Gladio and Ronaldo, respectively), whereas micronized wholemeal (M) showed a significant increase of 37% in Gladio and of 45% in Ronaldo (Figure 3B). This result could be due to the micronization process which produced a finer flour that could improve the fiber analytical determination, likely because of the increase in the surface area available for the enzyme activity, as already observed about starch [36]. Similar results were reported in micronized wheat [44] and barley flour [45]. The air fractionation process significantly increased the fiber content in both cultivars only in the F fraction (+63% in Gladio and +31% in Ronaldo, Figure 3B), supporting the air fractionation as an eligible technology, able to obtain flour fractions enriched in fibers [22,45] and indicated to improve the nutritional value of rice refined flours that contain low fiber content. However, the air fractionation process was affected by the genotype; indeed, in cv Gladio the major amount of TDF was present in the F fraction, while in cv Ronaldo, fiber was more equally distributed between the two milling fractions (Figure 3B). This result could be due to the different amylose and resistant starch content of the two cultivars which probably affected the textural properties of rice kernel and consequently the particle size of micronized flour.

In both cultivars, the parboiling process caused a small and not significant reduction in TAC levels in brown rice kernels (Figure 3C) because of a certain loss of compounds with antioxidant activity that are sensible to hydrothermal conditions [46]. A further significant decrease was observed in the refined flour (Figure 3C) in which the external layers, where the antioxidant compounds are mostly concentrated, are absent. Micronized wholemeal (M) and parboiled brown rice (PBR) showed similar and not significant differences in TAC levels (Figure 3C) in cv Gladio, whereas a slight but significant increase was observed in cv Ronaldo. As above observed for fiber, proteins, and resistant starch, F fraction showed the highest TAC level (Figure 3C), likely due to the presence of a major content of phenolic acids in the bran [47], indicating this fraction as the richest in bioactive and antioxidant compounds, and therefore, it has been selected to improve the nutritional potential of dry rice pasta. The addition of 15% of F fraction has been valuated as the best compromise in terms of nutritional and sensory texture aspects since this percentage is allowed to reach about 4% of fiber content in pasta formulations, so that they can be defined as a 'source of fiber'.

3.2. *Chemical Characterization and Color of Dry Pasta*

Since the chemical characteristics of the raw materials include a high level of fiber, resistant starch, TAC and proteins and low total starch content, two pasta samples for each rice cultivar were made, one from 100% micronized wholemeal (MP) and the other from refined rice flour replaced with 15% of the F fraction (RFP). These pasta formulations represented a *unicum* amongst enriched pasta, being obtained by non-conventional technological processes and by enrichment with fractions derived from rice cultivar itself.

Micronized pasta (MP) showed a total starch content slightly but not significantly lower than RFP in both cvs Gladio and Ronaldo (Table 1), due to the presence of the refined flour which mostly contributed (85%) to the RFP pasta formulation. In both cultivars, amylose, and RS content did not statistically differ in MP or in RFP (Table 1). However, it is noteworthy that both pasta samples from cv Gladio exhibited amylose and resistant starch values definitely higher than those revealed in MP and RFP from cv Ronaldo. As previously discussed for raw materials, these results confirmed the positive correlation between amylose and resistant starch content [35–38].

Statistically significant differences in protein content between MP and RFP pasta were observed only in cv Gladio, being one percentage point lower in RFP than in MP (Table 1). In RFP, the protein content decrease was not as high as expected thanks to the addition of 15% of the F fraction whose protein content was the highest amongst the raw materials (Figure 3A).

Gladio and Ronaldo MP, showed higher fiber content than RFP pasta (+38% and +13%, respectively; Table 1). Nevertheless, the replacement of refined flour with 15% of F fraction allowed to obtain a TDF content as high as 4.5% on average, very similar to the fiber content, 3.0–4.0% usually detected in durum semolina pasta [48].

Both MP and RFP evidenced a low but significant decrease in TAC levels compared to their relative starting raw materials (Table 1); likely the decrement in pasta samples could be explained by the rearrangements occurring in the pasta structure as a consequence of extrusion and drying processes which could affect the accessibility of the ABTS radical to the antioxidant compounds and their thermal degradation, as previously observed in durum wheat pasta by Martini et al. [49]. Noteworthy, in both cultivars, TAC values in RFP samples were very similar (Table 1) to those observed in MP ($p > 0.05$), indicating that the addition of only 15% of F fraction to refined flour determined a remarkable increment of TAC also in refined pasta sample.

Table 1. Basic composition and total antioxidant capacity of pasta.

	TS	Amylose	RS	Protein	TDF	TAC	Ash
	(%)	(%)	(%)	(%)	(%)	(mmol TEAC/kg)	(%)
Gladio MP	83 ± 0.7 [ab]	18.3 ± 0.9 [a]	0.7 ± 0.03 [a]	8.86 ± 0.06 [b]	6.2 ± 0.1 [a]	21.7 ± 0.3 [c]	1.59 ± 0.01 [a]
Gladio RFP	84.7 ± 0.3 [a]	17.8 ± 0.4 [a]	0.78 ± 0.03 [a]	7.73 ± 0.04 [c]	4.49 ± 0.04 [b]	22.0 ± 0.2 [c]	1.29 ± 0.01 [bc]
Ronaldo MP	84.2 ± 0.6 [a]	12.8 ± 0.4 [b]	0.22 ± 0.03 [c]	8.94 ± 0.07 [b]	5.2 ± 0.1 [ab]	24.4 ± 0.1 [b]	1.47 ± 0.01 [b]
Ronaldo RFP	85.9 ± 0.6 [a]	12.2 ± 0.4 [b]	0.19 ± 0.03 [c]	8.87 ± 0.04 [b]	4.6 ± 0.1 [b]	23.6 ± 0.3 [bc]	1.36 ± 0.01 [b]
Semolina pasta	78.3 ± 0.3 [b]	19.6 ± 0.7 [a]	0.38 ± 0.01 [b]	11.55 ± 0.03 [a]	4.2 ± 0.3 [b]	46.8 ± 0.6 [a]	0.71 ± 0.01 [c]

Results are expressed as mean ± standard deviation. Values with different letters within the same column indicate significant differences determined by the pairwise Mann–Whitney test ($p \leq 0.05$). MP = Micronized Pasta; RFP = Refined + F Fraction Pasta; TS = total starch; RS = resistant starch; TDF = total dietary fiber; TAC = total antioxidant capacity; TEAC = trolox equivalent antioxidant capacity.

As expected in MP, the ash content was higher than in RFP samples (Table 1), even considering that the F fraction gave a great contribution to ash content in RFP value. However, the ash content value, in all four pasta samples, fell within the Italian legal limit for whole semolina pasta (1.8%) [50].

Both pasta formulations from cv Ronaldo showed higher ($p \leq 0.05$) yellow index (b*) than pasta from cv Gladio (Table 2). Anyway, b* values in all pasta samples fell within the range 23–26 considered as 'good' for semolina pasta [51]. These good yellow indices could be mainly due to the parboiling process that allows enhancement of the yellowness of the milled rice kernels because of Maillard reactions and physico-chemical changes in starch and protein components [42,43,52,53]. Hence, all the rice spaghetti produced in the present study, presented good yellow indices, making these pasta formulations inviting even for consumers of durum semolina pasta. Brown (100-L) indices were very similar in all pasta samples, Gladio MP, being the highest (Table 2). However, brown and red values were higher ($p \leq 0.05$) than those usually obtained from semolina pasta because of the outer layers present in micronized wholemeal pasta (MP) and of the fiber-rich F fraction in RFP, but also as an effect of the parboiling process which is reported to influence brown and red color parameters [53].

Table 2. Color and cooking properties of rice and semolina pasta.

	Yellow index (b*)	Brown Index (100-L)	Red Index (a*)	OCT (min′ s″)	WA (g)	Cooking Loss (%)
Gladio MP	23.5 ± 0.4 [c]	58.4 ± 0.9 [a]	5.33 ± 0.07 [bc]	8′50″ ± 5″ [b]	54.9 ± 0.2 [d]	2.40 ± 0.01 [b]
Gladio RFP	24.3 ± 0.4 [c]	55.3 ± 0.2 [b]	4.5 ± 0.1 [d]	8′15″ ± 5″ [c]	96.9 ± 0.2 [c]	2.42 ± 0.02 [b]
Ronaldo MP	25.8 ± 0.3 [b]	56 ± 1 [b]	5.7 ± 0.3 [b]	8′50″ ± 5″ [b]	55.6 ± 0.2 [d]	3.096 ± 0.016 [a]
Ronaldo RFP	25.9 ± 0.3 [b]	55.8 ± 0.4 [b]	6.0 ± 0.1 [a]	8′15″ ± 5″ [c]	105.9 ± 0.3 [b]	2.29 ± 0.02 [c]
Semolina pasta	40.8 ± 0.2 [a]	41.5 ± 0.4 [c]	1.47 ± 0.08 [e]	10′30″ ± 5″ [a]	148.6 ± 0.5 [a]	0.367 ± 0.003 [d]

Results are expressed as mean ± standard deviation. Values with different letters within the same column indicate significant differences determined by the pairwise Mann–Whitney test ($p \leq 0.05$). MP = Micronized Pasta; RFP = Refined + F Fraction Pasta; OCT = optimal cooking time; WA = water absorption.

3.3. Cooking Quality

Rice pasta samples were compared to durum semolina pasta produced in the same pilot plant at the same extrusion and drying conditions. In both rice cultivars, the optimal cooking time (OCT) was significantly higher in MP, than in RFP (Table 2). The absence of gluten and the presence of the fiber might make the water absorption easier, since fiber has higher water absorption than gluten proteins, hence reducing the cooking time with respect to traditional semolina pasta. Indeed, a significant decrement was observed for water absorption (WA) in all pasta samples in comparison to semolina pasta (Table 2). Moreover, in both cultivars, MP adsorbed almost half the amount of water with respect to RFP. These results are in accordance with findings already observed in other bran-enriched pasta [27,54,55], since the fiber retains less water with respect to the starch. Through the parboiling process, this was demonstrated to induce lipid-amylose complex reducing starch swelling and amylose leaching during cooking [13], the absence of gluten caused, in all rice pasta formulations, a cooking loss heavy higher than in semolina pasta (Table 2), as already observed by Kaur et al. [56]. The global sensorial judgment (Figure 4), focused on the sensory texture quality traits, revealed that Gladio was the most suitable cultivar for rice pasta formulation, the RFP reaching values as good as those of semolina pasta, with the consensus of all of the three experienced assessors. Noteworthy, in both cultivars, the enrichment with F fraction improved both stickiness and bulkiness sensorial parameters whereas no effect was observed on firmness. This last parameter was instead mostly affected by the amylose (r^2 = 0.91) and resistant starch content (r^2 = 0.80) [14], which resulted higher in cv Gladio, hence improving rice spaghetti firmness. The observed huge differences between pasta from cv Gladio and cv Ronaldo highlighted the importance of the genotype choice to obtain products with suitable characteristics that could meet the rice pasta consumers' acceptance.

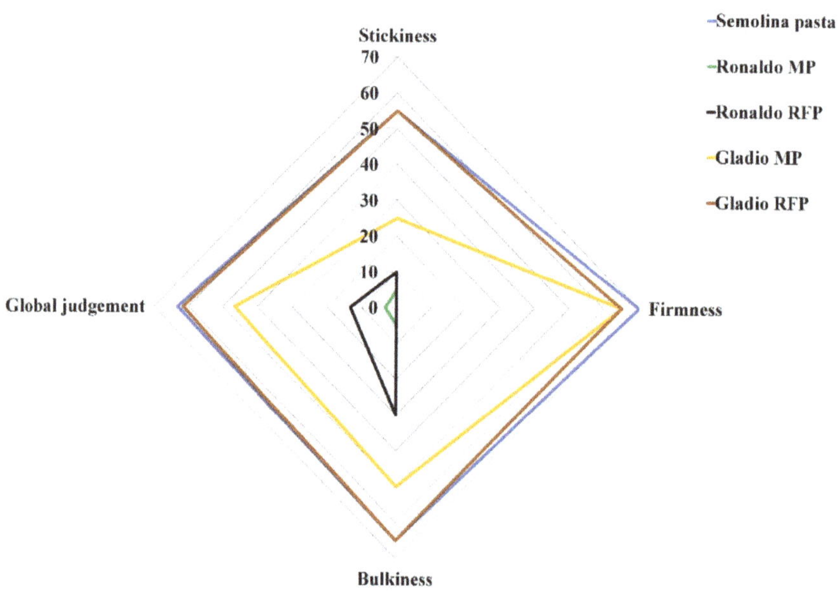

Figure 4. Radar chart of sensory assessment of rice and semolina pasta. MP = Micronized Pasta; RFP = Refined + F Fraction Pasta. For stickiness and bulkiness: ≤20 = very high, >20 and ≤40 = high, >40 and ≤60 = rare, >60 and ≤80 = almost absent, >80 and ≤100 = absent; for firmness, ≤20 = absent, >20 and ≤40 = rare, >40 and ≤60 = sufficient, >60 and ≤80 = good, >80 and ≤100 = very good. Global Sensorial Judgement score ranges from 10 to 100: <55 = scarce, ≥55 and <65 = sufficient, ≥65 and <75 = good, ≥75 = very good.

3.4. Starch Hydrolysis and Predicted Glycemic Index

Results obtained from the in vitro method for the determination of starch hydrolysis revealed that all rice pasta presented hydrolysis indices (HI) lower ($p \leq 0.05$) than semolina pasta, high amylose Gladio pasta samples showing lower ($p \leq 0.05$) values than Ronaldo ones (Table 3). It could be inferred that the parboiling process coupled with the presence of fiber, exerted a crucial role in lowering the starch hydrolysis and consequently the glycemic index (GI). Indeed, as already reported by Zohoun et al. [16], starch hydrolysis is more efficient for low than for high amylose rice cultivars resulting in a lower glycemic response. Moreover, the formation of resistant starch upon rice parboiling, reduced starch digestibility and glycemic index [57], because of crystallization of amylose after hydrothermal treatment, which changes starch accessibility by hydrolytic enzymes. The differences in HI and GI observed between the two varieties, indicated that the employment of a specific genotype [58] coupled with an appropriate technological process, could allow to obtain rice pasta with lower glucose release.

Table 3. Hydrolysis and predicted glycemic index of rice and semolina pasta.

	HI	GI
Gladio MP	46.10.4 ^{d}	65.00.2 ^{c}
Gladio RFP	471 ^{d}	65.30.6 ^{c}
Ronaldo MP	591.08 ^{c}	72.20.6 ^{b}
Ronaldo RFP	703 ^{b}	782 ^{ab}
Semolina pasta	82.00.6 ^{a}	84.70.3 ^{a}

Results are expressed as mean ± standard deviation. Values with different letters within the same column indicate significant difference determined by the pairwise Mann–Whitney test ($p \leq 0.05$). MP = Micronized Pasta; RFP = Refined + F Fraction Pasta; HI = hydrolysis index; GI = glycemic index.

Summing up, the biplot of PCA (Figure 5), obtained by combining all pasta sample variables, allowed to distinguish three main groups, Gladio and Ronaldo rice pasta in the second and third quadrants, respectively, whereas semolina pasta in the fourth one. Moreover, rice pasta occurred in two subpopulations attributed to the different genotypes which were discriminated in relation to amylose and resistant starch as well as in the three sensory texture parameters (bulkiness, stickiness, and firmness). The first two principal components (PC1 and PC2, Figure 5) accounted for 63.9% and 26.5% of the total variance, respectively. The first component was positively associated mainly with protein content, TAC level, b* value, OCT, and WA and negatively with ash and TS content, CL, brown and red indices. The second component was mostly associated with RS and amylose content, and sensory texture parameters and negatively with HI and GI.

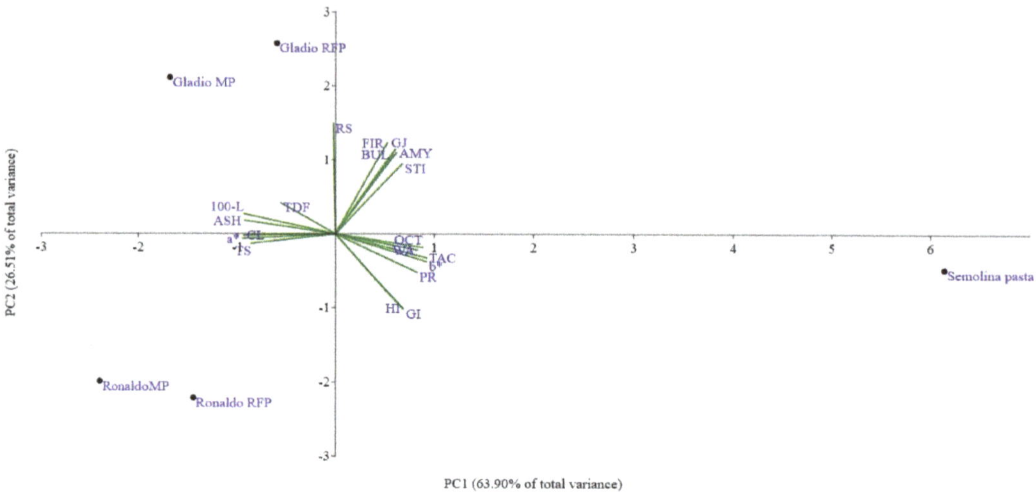

Figure 5. PCA biplot of descriptive analysis of sensory texture parameters: firmness (FIR), bulkiness (BUL), stickiness (STI), global sensorial judgment (GJ); of technological and nutritional parameters: yellow (b*), brown (100-L) and red (a*) indices, water absorption (WA), cooking loss (CL), optimal cooking time (OCT), ash (ASH), total starch (TS), resistant starch (RS), amylose (AMY), protein content (PR), total dietary fiber (TDF), total antioxidant capacity (TAC), hydrolysis (HI), and glycemic (GI) indices, detected in semolina pasta, in micronized pasta (MP) and in refined + F fraction pasta (RFP) from rice cvs Gladio and Ronaldo.

4. Conclusions

In this study, two rice cultivars differing in amylose content were evaluated to produce brown rice spaghetti obtained from parboiled micronized or refined flour-enriched with air classified fine fraction. The addition of this flour fraction allowed to obtain rice pasta (RFP) with superior sensorial performances than wholegrain pasta (MP), with comparable nutritional properties, by the exclusive use of rice itself. Besides the technological processes, pasta quality was affected the most by the genotype, since pasta obtained from high amylose cv Gladio resulted in the best in terms of nutritional, technological, and sensorial quality. Indeed, pasta from Gladio showed the highest RS content and the lowest glycemic index along with sensorial characteristics as good as durum semolina pasta. Further studies should be addressed to investigate the behavior of additional rice cultivars with contrasting amylose content in terms of pasta making attitude, in order to confirm the role of amylose and resistant starch in pasta quality. Finally, in vivo test should be conducted with the aim to estimate the post-prandial blood glucose content upon the ingestion of differently processed rice pasta products.

Author Contributions: Conceptualization, F.T. and L.G.; methodology, F.N., F.T. and E.G.; validation, F.N., F.T., E.G. and L.G.; formal analysis, F.T. and E.G., investigation, F.T. and E.G., resources, L.G.; data curation, F.N. and L.G.; writing—original draft preparation, F.N. and L.G.; writing—review and editing, F.N., F.T., E.G. and L.G.; visualization, E.G. and F.T.; supervision, L.G.; project administration, L.G.; funding acquisition, L.G. All authors have read and agreed to the published version of the manuscript.

Funding: This research was funded by MIPAAF, DM 5237 05/12/2011 (POLORISO).

Institutional Review Board Statement: Not applicable.

Informed Consent Statement: Not applicable.

Data Availability Statement: Data are contained within the article.

Acknowledgments: The Authors would like to thank Roberto Mortaro, Ester Gosparini and Alessandra Arcangeli of CREA-IT of Rome for their technical assistance in pasta making process and sensorial judgment and D.ssa Daniela Sgrulletta as former Director of the project.

Conflicts of Interest: The authors declare no conflict of interest. The funders had no role in the design of the study; in the collection, analyses, or interpretation of data; in the writing of the manuscript, or in the decision to publish the results.

References

1. Durand-Morat, A.; Nalley, L.L.; Thoma, G. The implications of red rice on food security. *Glob. Food Sec.* **2018**, *18*, 62–75. [CrossRef]
2. Verma, D.K.; Srivastav, P.P. Bioactive compounds of rice (*Oriza sativa* L.): Review on paradigm and its potential benefit in human health. *Trends Food Sci. Technol* **2020**, *97*, 355–365. [CrossRef]
3. Niland, B.; Cash, B.D. Health benefits and adverse effects of a gluten-free diet in non-celiac disease patients. *Gastr. Hepatol.* **2018**, *14*, 82–91.
4. Deshmukh, R.; Thomas, A. Allied Market Research/Gluten Free Pasta Market Report. 2019. Available online: https://www.alliedmarketresearch.com/gluten-free-pasta-market (accessed on 19 March 2021).
5. Rosell, C.; Marco, C. Rice. In *Gluten Free Cereal Products and Beverages*; Aarendt, E., Dal Bello, F., Eds.; Academic Press: New York, NY, USA; Elsevier: London, UK, 2008; pp. 81–100.
6. Padalino, L.; Conte, A.; Del Nobile, M.A. Overview on the general approaches to improve gluten-free pasta and bread. *Foods* **2016**, *5*, 87. [CrossRef]
7. Marti, A.; Pagani, M.A. What can play the role of gluten in gluten free pasta? *Trends Food Sci. Technol.* **2013**, *31*, 63–71. [CrossRef]
8. Lee, S.; Lee, J.H.; Chung, H.J. Impact of diverse cultivars on molecular and crystalline structures of rice starch for food processing. *Carbohydr. Polym.* **2017**, *169*, 33–40. [CrossRef] [PubMed]
9. Wu, N.N.; Tan, B.; Li, S.S.; Zhang, M.; Tian, X.H.; Zhai, X.T.; Liu, M.; Liu, Y.X.; Wang, L.P.; Gao, K. Quality characteristics of extruded brown rice noodles with different amylose contents. *Food Sci. Technol. Res.* **2018**, *24*, 311–319. [CrossRef]
10. Panlasigui, L.N.; Thompson, L.U.; Juliano, B.O.; Perez, C.M.; Yiu, S.H.; Greenberg, G.R. Rice varieties with similar amylose content differ in starch digestibility and glycemic response in humans. *Am. J. Clin. Nutr.* **1991**, *54*, 871–877. [CrossRef]
11. Li, C.; You, Y.; Chen, D.; Gu, Z.; Zhang, Y.; Holler, T.P.; Ban, X.; Hong, Y.; Cheng, L.; Li, Z. A systematic review of rice noodles: Raw material, processing method and quality improvement. *Trends Food Sci. Technol.* **2021**, *107*, 389–400. [CrossRef]
12. Miao, M.; Jiang, B.; Zhang, T. Effect of pullulanase debranching and recrystallization on structure and digestibility of waxy maize starch. *Carbohydr. Polym.* **2009**, *76*, 214–221. [CrossRef]
13. Bhattacharya, K.R. Parboiling of rice. In *Rice Chemistry and Technology*, 3rd ed.; Champagne, E.T., Ed.; AACC Int. Press: St. Paul, MN, USA, 2004; pp. 329–404.
14. Marti, A.; Seetharaman, K.; Pagani, M.A. Rice-based pasta: A comparison between conventional pasta-making and extrusion-cooking. *J. Cereal Sci.* **2010**, *52*, 404–409. [CrossRef]
15. Kim, J.C.; Mullan, B.P.; Hampson, D.J.; Pluske, J.R. Effects of amylose content, autoclaving, parboiling, extrusion, and post-cooking treatments on resistant starch content of different rice cultivars. *Aust. J. Agric. Res.* **2006**, *57*, 1291–1296. [CrossRef]
16. Zohoun, E.V.; Ndindeng, S.A.; Soumanou, M.M.; Tang, E.N.; Bigoga, J.; Manful, J.; Sanyang, S.; Akissoe, N.H.; Futakuchi, K. Appropriate parboiling steaming time at atmospheric pressure and variety to produce rice with weak digestive properties. *Food Sci. Nutr.* **2018**, *6*, 757–764. [CrossRef]
17. Englyst, H.N.; Kingman, S.M.; Cummings, J.H. Classification and measurement of nutritionally important starch fractions. *Eur. J. Clin. Nutr.* **1992**, *46*, 33–50.
18. Nugent, A.P. Health properties of resistant starch. *Nutr. Bull.* **2005**, *30*, 27–54. [CrossRef]
19. Grugni, G.; Mazzini, F.; Viazzo, G.; Viazzo, N. Process for the Production of Rice Pasta and Product Obtained Thereby. European Patent EP2110026A1, 2009. Available online: https://patents.google.com/patent/EP2110026A1/en (accessed on 16 March 2021).
20. Marti, A.; Caramanico, R.; Bottega, G.; Pagani, M.A. Cooking behavior of rice pasta: Effect of thermal treatments and extrusion conditions. *LWT Food Sci. Technol.* **2013**, *54*, 229–235. [CrossRef]
21. Fukagawa, M.K.; Zsika, L.H. Rice: Importance for global nutrition. *J. Nutr. Sci. Vitaminol.* **2019**, *65*, S2–S3. [CrossRef]
22. Ciccoritti, R.; Nocente, F.; Sgrulletta, D.; Gazza, L. Cooking quality, biochemical and technological characteristics of bran-enriched pasta obtained by a novel pasta-making process. *LWT Food Sci. Technol.* **2019**, *101*, 10–16. [CrossRef]
23. Juliano, B.O. Structure chemistry and function of the rice grain and its fraction. *Cereal Foods World* **1992**, *37*, 772–774.
24. Hidalgo, A.; Brandolini, A.; Gazza, L. Influence of steaming treatment on chemical and technological characteristics of einkorn (*Triticum monococcum* L. ssp. *monococcum*) wholemeal flour. *Food Chem.* **2008**, *111*, 549–555. [CrossRef]
25. American Association of Cereal Chemists. Pasta and noodle cooking quality-firmness. In *Approved Methods of Analysis, 66-50.01*, 11th ed.; AACC International: St Paul, MN, USA, 2000.
26. D'Egidio, M.G.; Mariani, B.M.; Nardi, S.; Novaro, P.; Cubadda, R. Chemical and technological variables and their relationships: A predictive equation for pasta cooking quality. *Cereal Chem.* **1990**, *67*, 275–281.
27. Nocente, F.; Taddei, F.; Galassi, E.; Gazza, L. Upcycling of brewers' spent grain by production of dry pasta with higher nutritional potential. *LWT-Food Sci. Technol.* **2019**, *114*, 108421. [CrossRef]

28. Association of Official Analytical Chemists. *Official Methods of Analysis of AOAC International, Method 996.11*, 18th ed.; AOAC: Gaithersburg, MD, USA, 2005.
29. Association of Official Analytical Chemists. *Official Methods of Analysis of AOAC International, Method 2002.02*, 18th ed.; AOAC: Gaithersburg, MD, USA, 2005.
30. Association of Official Analytical Chemists. *Official Methods of Analysis 991*, 16th ed.; Cunniff, P., Ed.; AOAC: Gaithersburg, MD, USA, 1995; p. 42.
31. American Association of Cereal Chemists. *Approved Methods of Analysis, 08-01.01 Ash*; AACC International: St. Paul, MN, USA, 2013.
32. International Association for Cereal Science and Technology. *ICC Standard Methods (Methods No. 105/2)*; ICC: Vienna, Austria, 2003.
33. Martini, D.; Taddei, F.; Nicoletti, I.; Ciccoritti, R.; Corradini, D.; D'Egidio, M.G. Effects of genotype and environment on phenolic acids content and total antioxidant capacity in durum wheat. *Cereal Chem.* **2014**, *91*, 310–317. [CrossRef]
34. Goñi, I.; Garcia-Alonso, A.; Saura-Calixto, F. A starch hydrolysis procedure to estimate glycemic index. *Nutr. Res.* **1997**, *17*, 427–437. [CrossRef]
35. Biselli, C.; Volante, A.; Desiderio, F.; Tondelli, A.; Gianinetti, A.; Finocchiaro, F.; Taddei, F.; Gazza, L.; Sgrulletta, D.; Cattivelli, L.; et al. GWAS for starch-related parameters in japonica rice (*Oryza sativa* L.). *Plants* **2019**, *8*, 292. [CrossRef] [PubMed]
36. Hossen, M.S.; Sotome, I.; Takenake, M.; Isobe, S.; Nakajima, M.; Shimizu, N.; Okadome, H. Ultra-fine pulverization of rice: Effects on hydration properties and enzymatic hydrolysis. *Jpn. J. Food Eng.* **2013**, *14*, 37–46. [CrossRef]
37. Shu, X.; Sun, J.; Wu, D. Effects of grain development on formation of resistant starch in rice. *Food Chem.* **2014**, *164*, 89–97. [CrossRef]
38. Kong, X.; Chen, Y.; Zhu, P.; Sui, Z.; Corke, H.; Bao, J. Relationships among genetic, structural, and functional properties of rice starch. *J. Agric. Food Chem.* **2015**, *63*, 6241–6248. [CrossRef]
39. Marti, A.; Pagani, M.A.; Seetharaman, K. Understanding starch organization in gluten-free pasta from rice flour. *Carbohydr. Polym.* **2011**, *84*, 1069–1074. [CrossRef]
40. Wang, S.; Li, C.; Copeland, L.; Niu, Q.; Wang, S. Starch retrogradation: A comprehensive review. *Compr. Rev. Food Sci. Food Saf.* **2015**, *14*, 568–585. [CrossRef]
41. Otegbayo, B.O.; Osamuel, F.; Fashakin, J.B. Effect of parboiling on physico-chemical qualities of two local rice varieties in Nigeria. *J. Food Technol. Afr.* **2001**, *6*, 130–132. [CrossRef]
42. Pall, P.; Singh, N.; Kaur, P.; Kaur, A. Effect of parboiling on phenolic, protein, and pasting properties of rice from different paddy varieties. *J. Food Sci.* **2018**, *83*, 2761–2771. [CrossRef]
43. Zohoun, E.V.; Tang, E.N.; Soumanour, M.M.; Manfull, J.; Akisoe, N.H.; Bigoga, J.; Futakuchi, K.; Ndindeng, S.A. Physiochemical and nutritional properties of rice as affected by parboiling steaming time at atmospheric pressure and variety. *Food Sci. Nutr.* **2018**, *6*, 638–652. [CrossRef]
44. Rizzello, C.G.; Coda, R.; Mazzacane, F.; Minervini, D.; Gobbetti, M. Micronized by-products from debranned durum wheat and sourdough fermentation enhanced the nutritional, textural and sensory features of bread. *Food Res. Int.* **2012**, *46*, 304–313. [CrossRef]
45. Ferrari, B.; Finocchiaro, F.; Stanca, A.M.; Gianinetti, A. Optimization of air classification for the production of beta-glucan–enriched barley flours. *J. Cereal Sci.* **2009**, *50*, 152–158. [CrossRef]
46. Walter, M.; Marchesan, E.; Massoni, P.F.S.; da Silva, L.P.; Sartori, G.M.S.; Ferreira, R.B. Antioxidant properties of rice grains with light brown, red and black pericarp colors and the effect of processing. *Food Res. Int.* **2013**, *50*, 698–703. [CrossRef]
47. Zhou, Z.; Robards, K.; Helliwell, S.; Blanchard, C. The distribution of phenolic acids in rice. *Food Chem.* **2004**, *87*, 401–406. [CrossRef]
48. Shewry, P.R. The HEALTHGRAIN program opens new opportunities for improving wheat for nutrition and health. *Nutr. Bull.* **2009**, *34*, 225–231. [CrossRef]
49. Martini, D.; Ciccoritti, R.; Nicoletti, I.; Nocente, F.; Corradini, D.; D'Egidio, M.G.; Taddei, F. From seed to cooked pasta: Influence of traditional and non-conventional transformation processes on total antioxidant capacity and phenolic acid content. *Int. J. Food Sci. Nutr.* **2018**, *69*, 24–32. [CrossRef] [PubMed]
50. Italian Presidential Decree n. 187, 9 February 2001. Available online: https://www.politicheagricole.it/flex/cm/pages/ServeBLOB.php/L/IT/IDPagina/2772 (accessed on 23 February 2021).
51. *Prodotti Derivati dal Grano duro Destinati alla Pastificazione. Definizione, Caratteristiche e Classi di Qualità*; Uni Standard 10940:2001; UNI, Ente Nazionale Italiano di Unificazione: Milano, Italy, 2001.
52. Bhattacharya, S. Kinetics on colour changes in rice due to parboiling. *J. Food Eng.* **1996**, *29*, 99–106. [CrossRef]
53. Lamberts, L.; De Bie, E.; Derycke, V.; Veraverbeke, W.S.; De Mann, W.; Delcour, J.A. Effect of processing conditions on colour change of brown and milled parboiling rice. *Cereal Chem.* **2006**, *83*, 80–85. [CrossRef]
54. Aravind, N.; Sissons, M.; Egan, N.; Fellows, C.M. Effect of insoluble dietary fibre addition on technological, sensory, and structural properties of durum wheat spaghetti. *Food Chem.* **2012**, *130*, 299–309. [CrossRef]
55. Wang, L.; Duan, W.; Zhou, S.; Qian, H.; Zhang, H.; Qi, X. Effect of rice bran fibre on the quality of rice pasta. *Int. J. Food Sci. Technol.* **2018**, *53*, 81–87. [CrossRef]
56. Kaur, G.; Sharma, S.; Nagi, H.P.; Dar, B.N. Functional properties of pasta enriched with variable cereal brans. *J. Food Sci. Technol.* **2012**, *49*, 467–474. [CrossRef]

57. Barbiroli, A.; Bonomi, F.; Casiraghi, M.C.; Iametti, S.; Pagani, M.A.; Marti, A. Process conditions affect starch structure and its interactions with proteins in rice pasta. *Carbohydr. Polym.* **2013**, *92*, 1865–1872. [CrossRef]
58. Kumar, A.; Sahoo, U.; Baisakha, B.; Okpani, O.A.; Ngangkham, U.; Parameswaran, C.; Basak, N.; Kumar, G.; Sharma, S.G. Resistant starch could be decisive in determining the glycemic index of rice cultivars. *J. Cereal Sci.* **2018**, *79*, 348–353. [CrossRef]

foods

MDPI

Article

Innovative Development of Pasta with the Addition of Fish By-Products from Two Species

Andrea Ainsa, Adrián Honrado, Pedro L. Marquina, Pedro Roncalés, José Antonio Beltrán and Juan B. Calanche M. *

Agroalimentario de Aragón-IA2, Universidad de Zaragoza-CITA, Miguel Servet, 177, 50013 Zaragoza, Spain; andreaainsa6@gmail.com (A.A.); adrihonfri@gmail.com (A.H.); pmarquin@gmail.com (P.L.M.); roncales@unizar.es (P.R.); jbeltran@unizar.es (J.A.B.)

* Correspondence: calanche@unizar.es

Abstract: The fish industry generates by-products that are still nutrient-rich. Its incorporation in pasta production could be an interesting option to get functional food. Therefore, the aim of this study was to compare the nutritional composition, technological properties and sensory quality of two pastas containing tuna and sea bass by-products, separately. Durum wheat semolina and fish by-product concentrates were used in pasta manufacturing. Fatty acids profile, optimal cooking time, texture profile analysis, color, weight gain, swelling index, cooking losses and moisture were determined and compared with a non-containing fish reference. A sensory analysis was also carried out. In general, results showed a higher content of fatty acids in tuna pasta than in sea bass pasta. The texture profile analysis (TPA) showed lower hardness and fracturability in the fish pasta. Cohesiveness was higher in the tuna pasta while sea bass pasta was brighter. Fish incorporation caused a decrease in weight gain and swelling index and an increase in cooking losses. Sensory analysis established differences in homogeneity, typical aroma, fish flavor, fish odor and elasticity. It was concluded that the use of these by-products results in a more nutritious pasta although tuna content should be reduced (<3%) to improve its sensory profile.

Keywords: enriched pasta; bioactive compound; tuna; sea bass; Ω-3 fatty acids; sensometrics; TPA (texture profile analysis)

Citation: Ainsa, A.; Honrado, A.; Marquina, P.L.; Roncalés, P.; Beltrán, J.A.; Calanche M., J.B. Innovative Development of Pasta with the Addition of Fish By-Products from Two Species. *Foods* **2021**, *10*, 1889. https://doi.org/10.3390/foods10081889

Academic Editors: Laura Gazza and Francesca Nocente

Received: 20 July 2021
Accepted: 13 August 2021
Published: 15 August 2021

1. Introduction

An essential component of modern society is the consumption of functional foodstuffs and nutraceuticals [1]. Fundamentally, the main purpose among consumers to intake these foods is the aim to reduce the risk of chronic diseases or to enhance health [2]. The biological effect obtained from functional foodstuffs is related to different compounds which are naturally present or intentionally added to the product. These products can be classified as fortified, increasing the content of a natural component, or enriched, adding an external component. In this sense, important nutritional components are incorporated in different foods, such as Ω-3 fatty acids, amino acids (e.g., arginine, leucine, and tyrosine), and minerals (calcium, phosphorus or manganese among others) [3].

Pasta seems to be an excellent opportunity to be enriched by incorporating alternative bioactive compounds as it is a common food due to its easy handling, storage, and preparation, in addition to its low cost [4,5]. Considering that fish contains great values of Ω-3 fatty acids, pasta enriched with fish would be a good chance to achieve the suggested daily intake of healthier fatty acids and it could be defined as high content of Ω-3 fatty acids according to Council Regulation (CE) N° 116/2010 [6–8]. Fish has a high content of protein, essential amino acids, and it is a great source of vitamins (A, D, B6 and B12) and a wide variety of minerals (phosphorus, magnesium, iron, zinc, and iodine) [9,10].

The energy value of fish is largely determined by their content of lipids. This fat is a relevant source of bioactive compounds. The evaluation of farmed sea bass by-products indicates that it is feasible for the use of this fish to obtain MUFA, PUFA, Ω-3 fatty acids, minerals, proteins and amino acids [11]. However, blue fish contains more Ω-3 fatty acids, DHA and EPA than white fish [12]. For this reason, tuna (*Thunnus obesus*) from fishing is an excellent source of high-quality protein and Ω-3 polyunsaturated fatty acids [13].

Several researchers have evaluated the importance and potential use of food industry by-products [14,15], which could also reduce the environmental impact of this industrial sector. The development of strategies to take advantage of food industry by-products is a necessary action to improve the efficiency of industrial operations, reduce waste, and recover high-added value compounds for further utilization in food products which is named as circular economy [16]. The action plan of circular economy established measures covering the whole life cycle: from production and consumption to waste management and the market for secondary raw materials and a revised legislative proposal on waste [17]. It is intended to achieve some of the sustainable development goals among which are: (2) zero hunger, (3) good health and well-being, (12) responsible consumption and production or (14) life below water [18].

The present study was carried out to compare the effect of two different fish by-products from a distinct origin (fished and farmed) in pasta-making, assessing their nutritional values, technological properties, and carrying out a sensory study. The main purpose was to achieve a great enrichment in ALA, EPA and DHA through the addition of fish.

2. Materials and Methods

2.1. Raw Material

Sea bass (*Dicentrarchus labrax*) by-products (flawed fillets and flesh cut) from aquaculture were supplied by a local fish industry (Scanfisk®, Zaragoza, Spain) and tuna (*Thunnus obesus*) by-products (head and flesh cut) were supplied by the local fishermen of Pontevedra, Spain. With respect to the cereal source, semolina from durum wheat (*Triticum durum*) was provided by a local company (Pastas Romero®, Zaragoza, Spain). The antioxidant used was rosemary extract powder (E-392) provided by Marbys® (Barcelona, Spain).

2.2. Enriched Pasta Preparation

The concentrates (≈18.5% moisture) were produced in the same way. Frozen fillets and cuts from sea bass and tuna (by-products) were manually deboned meat (MDM) and processed according to the methodology described in our previous studies [19,20]. Two types of pasta with durum wheat semolina (≈11.7% moisture) were made from each species used. They were produced with an experimental extrusion machine (Imperia & Monferrina, Mod. P6 LM14040, Moncalieri, Italy) in *fusilli* format according to Calanche et al., 2019. Enriched pasta with both fish concentrates was desiccated to obtain dry pasta (≈10% moisture) that was stored at room temperature. A control pasta composed of durum wheat and water was manufactured as a comparison for some of the analyses. Pasta formulations and their proximal analyses are shown in Table 1.

Table 1. Formulations used to make enriched pasta with fish by-products and their proximal analyses.

Ingredients	Sea Bass Pasta (%)	Tuna Pasta (%)	Control Pasta (%)
Durum wheat	72	72	75
Dried fish concentrate	3	3	0
Water	25	25	25
Proximal Composition (%)	**Sea Bass Pasta (%)**	**Tuna Pasta (%)**	**Control Pasta (%)**
Moisture	10.6	10.4	11.6
Protein	14.8	13.2	12.5
Fat	1.5	1.8	1.4
Fiber	1.23	1.27	1.20

2.3. Fatty Acids Profile

Fatty acids profile was determined according to Bligh and Dyer method (1959) [19], taking into account modifications made in the analysis protocol in agreement with the procedure described by Ainsa et al. (2021) [20] to get a better adjustment to assayed enriched pasta. Each sample was homogenized with different solvents (chloroform, methanol, potassium chloride and water) using an Ultraturrax device (IKA-WERKE, T-25 basic). Subsequently, it was centrifuged at 4000 rpm at 10 min, and the fat was extracted. Solvents were evaporated with BHT (butylated hydroxytoluene) as an antioxidant. Then, 2 mL of hexane and 1 mL of potassium hydroxide saturated were incorporated. Fatty acid profile was analyzed using a gas chromatograph (HP-6890II). Fatty acids were measured as the total area (%) of identified fatty acids.

2.4. Optimal Cooking Time

The optimal cooking time was estimated with a Warner–Bratzler cut test according to the instruction manual of the texturometer used (ANAME, TA-XT2i). The determination of the optimal cooking time was made using the texturometer with a flat Warner–Bratzler device. The instrumental measure of hardness was carried out in cooked pasta according to sample times assayed. Hardness was defined as the maximum force (tangential angle) required to cut the sample and was expressed in kg/mm^*s^2. Test conditions development was: pre-test speed: 2 mm/s; test speed: 2 mm/s; post-test speed: 10 mm/s; cutting distance: 15 mm; threshold strength: 0.010 kg. The hardness of pasta was determined by triplicate.

2.5. Texture Profile Analysis—TPA

A texturometer (ANAME, TA-XT2i) with a cylindrical flat aluminum probe was used for texture profile analysis (TPA). The method consisted of the application of two compression cycles with decompression of 20 s over cooked pasta. In this way, it was possible to determine hardness, adhesiveness, springiness, cohesiveness, gumminess, chewiness and fracturability. The conditions were: test speed: 2 mm/s; sample deformation: 75%; force threshold: 10 g. Five measurements were made for each type of pasta.

2.6. Pasta Color

Color analysis in cooking pasta according to its optimal cooking time was made using a colorimeter (Minolta. CM-2002, Japan). The CIE L*a*b* system represented by L* (brightness), a* (redness) and b* (yellowness) was used. The color variation produced by each fish species was calculated with a total color difference (ΔE) between control pasta and sea bass and tuna pasta:

$$\Delta E = \sqrt{(\Delta L^*)^2 + (\Delta a^*)^2 + (\Delta b^*)^2} \tag{1}$$

ΔL* = L* Fish pasta- L* Control pasta; Δa* = a* Fish pasta- a* Control pasta and

Δb* = b* Fish pasta- b* Control pasta

2.7. Technological Properties

2.7.1. Weight Gain and Swelling Index

The weight gain was established with 3 g of pasta which was cooked in 180 mL of water during optimal cooking time, they were cooled in 100 mL of water; then, pasta was dried superficially with absorbent paper and weighed on an analytical balance [21]. This parameter was calculated from the following formula:

$$\text{WG}(\%) = \frac{Weight\ of\ cooked\ pasta - Weight\ of\ cooked\ pasta\ after\ drying}{Weight\ of\ cooked\ pasta\ after\ drying} \cdot 100 \tag{2}$$

Then, the cooked pasta was dehydrated in an oven at 105 °C for 24 h. The swelling index was determined by the following equation:

$$SI = \frac{Weight\ of\ cooked\ pasta\ (g)}{Weight\ of\ dried\ cooked\ pasta\ (g)} \quad (3)$$

2.7.2. Cooking Losses

A sample of 3 g of each pasta was added in 180 mL of water and cooking during the optimal cooking time [22]. The water resulting after cooking was collected in crucibles and evaporated on a stove at 105 °C until reaching a constant weight. The residue was weighed and determined as a percentage of the total weight of raw pasta.

2.7.3. Moisture

Moisture was evaluated by a gravimetric method. The pasta was weighed and then dried in an oven at 105 °C until reaching a constant weight. It was cooled to room temperature and weighed again.

$$Moisture\ (\%) = \frac{raw\ pasta\ weight - dried\ pasta\ weight}{raw\ pasta\ weight} \cdot 100 \quad (4)$$

2.8. Sensory Analysis

A panel of ten selected assessors with previous experience in sensory analysis of fish and pasta bellowing to the staff of Meat Science and Technology Official Research Group (A04_20R DGA) from the University of Zaragoza was used to carry a sensory method knowing as "deviation respect to a reference" -DR- [23] which uses a reference sample (control food) against all evaluated samples (assessed food). The assessors had demonstrated sensory sensitivity in preliminary tests, received considerable training and they were able to make consistent and repeatable sensory assessments of various samples of pasta. The panel received prior training with respect to the use of the DR method and intensity scales to evaluate different attributes in pasta according to requirements of ISO standards (ISO 8586: 2012) [24]. Along this process, panelists became familiarized with the different descriptors and their intensity scales in order to assess the samples in a more accurate form [25]. The attributes selected for this study were: homogeneity, characteristic color, typical aroma, fish odor, rancidity flavor, hardness, elasticity, pastiness, pasta characteristic flavor, fish flavor and after-taste. All of them are based on previous studies [19,20].

Once the panel was prepared, the trained assessors indicated the degree of difference in intensity for each sensory attribute using a non-structured lineal scale of 10 cm anchored to the extremes as "none" to "much". The pasta was prepared by boiling until the optimal cooking time previously established and was served without any accompaniment at 60 °C according to Standard UNE-ISO 6658:2019 [23]. Each enriched pasta was served in an independent trial together with the reference pasta (control) and both were evaluated at the same time.

2.9. Statistical Analyses

Results of this study were analyzed using an XLSTAT Version 2016 (Addinsoft©, Paris, France). A univariate analysis was performed to check the normality of the data and detect outliers. Then, statistical analysis was performed by simple ANOVA (types of pasta) and Fisher test with a 95% confidence interval was used a posteriori to find differences among means for physical and chemical measures. To get a comparison between TPA and fatty acids content, a Pearson correlation was made and then, principal component analysis (PCA) was performed to explore relationships or associations that were of interest for this set of variables. In the sensory analysis, panel analyses were performed to establish the reliability of the results, verifying the panel's performance as well as its discriminative power. Posteriorly, ANOVA was performed to obtain significant differences with respect to control using a Dunnett test *a posteriori* (95% confidence interval) to establish differences between each type of pasta and control pasta. Then, the Fisher test was performed to find differences among all pasta (control/tuna/sea bass). Finally, characterization of each pasta was made with the square cosine method to get sensorial profiles, which were represented in a biplot where confidence ellipses of Hoteling (95%) were drawn to compare the samples.

3. Results

3.1. Comparison of Fatty Acids Profiles

The fatty acid profiles for both types of pasta is shown in Table 2.

Table 2. Fatty acid profiles for enriched pasta with fish by-products.

Fatty Acids	Tuna Pasta	Sea Bass Pasta
C14	2.44 ± 0.09 b	0.96 ± 0.01 a
C15	0.53 ± 0.02 b	0.06 ± 0.09 a
C16	21.31 ± 0.10 b	18.04 ± 0.21a
C17	0.56 ± 0.01 b	0.13 ± 0.11a
C18	3.08 ± 0.09 b	2.65 ± 0.15 a
C20	0.38 ± 0.01 b	0.05 ± 0.09 a
C22	0.18 ± 0.00 b	0.04 ± 0.07 a
%SFA	28.48 b	21.93 a
C16:1	2.81 ± 0.40 b	1.49 ± 0.02 a
C17:1	0.31 ± 0.01	0.41 ± 0.36
tC18:1 n-9	0.05 ± 0.09	0.05 ± 0.09
C18:1 n-11	1.85 ± 0.03 b	1.60 ± 0.04 a
C18:1 n-9 (oleic)	12.82 ± 0.05 a	20.75 ± 0.57 b
C20:1	2.09 ± 0.28	2.01 ± 0.70
C22:1 n-9	0.00 ± 0.00	0.16 ± 0.04
C24:1	0.12 ± 0.10	0.00 ± 0.00
%MUFA	20.04 a	26.47 b
C18:3 Ω-3 (ALA)	0.57 ± 0.03 a	2.98 ± 0.68 b
tC18:2 n-6	0.00 ± 0.00	0.05 ± 0.04
C18:2 n-6 (linoleic)	38.00 ± 0.38 a	43.97 ± 0.09 b
C20:2 n-6	0.30 ± 0.18	0.48 ± 0.02
C20:3 n-6	0.07 ± 0.06	0.00 ± 0.00
C22:2 n-6	0.09 ± 0.08	0.04 ± 0.02
C20:4 n-6	0.59 ± 0.02 b	0.15 ± 0.12 a
C22:6 Ω-3 (DHA)	7.31 ± 0.23 b	2.05 ± 0.12 a
C20:5 Ω-3 (EPA)	3.64 ± 0.10 b	1.44 ± 0.05 a
C22:5 Ω-3	0.43 ± 0.01	0.44 ± 0.02
%PUFA	50.43	51.59
ΣΩ3	11.37 b	6.91 a
ΣΩ6	39.06 a	44.69 b
P/S ratio	1.77 a	2.35 b
Ω6/Ω3 ratio	3.43 a	6.47 b
mg Ω3/100 g	202.42 b	111.88 a
%DRI (EFSA)	80.97 b	44.75 a

SFA: Saturated Fatty Acids, MUFA: Monounsaturated fatty acids, PUFA: polyunsaturated fatty acids, P/S ratio: PUFA/SFA ratio, %DRI: dietary reference intake. Lowercase letters show significant differences between both types of pasta ($p < 0.05$).

Saturated fatty acids percentage was significantly higher in tuna pasta than in seabass pasta. However, monounsaturated fatty acids were higher for sea bass pasta, especially due to oleic acid. Concerning polyunsaturated fatty acids, there were no significant differences between both percentages although EPA and DHA contents were higher in tuna pasta. The values of Ω3 and Ω6 presented a contrary behavior, while Ω3 value was higher in tuna pasta, Ω6 was higher in sea bass due to the content of linoleic acid.

Regarding fatty acids ratios, P/S and Ω6/Ω3 ratios were significantly higher in pasta enriched with sea bass. Nevertheless, the Ω3 content (mg/100 g) in pasta with tuna was almost double that in pasta with sea bass, thus, it presented a significant difference. The percentage of DRI (dietary reference intake of Ω3) was higher in tuna pasta because of the Ω3 content.

3.2. *Optimal Cooking Time*

The behavior about optimal cooking time for tuna and sea bass pasta was similar. Hardness reached an inflection point which is considered to be the optimal cooking time. Therefore, the perfect time for cooking was 210 s for pasta with fish while for control pasta was 390 s due to the significant decrease in hardness as shown in Figure 1. Although the time was the same for both kinds of pasta with fish added, tuna pasta was harder than sea bass pasta.

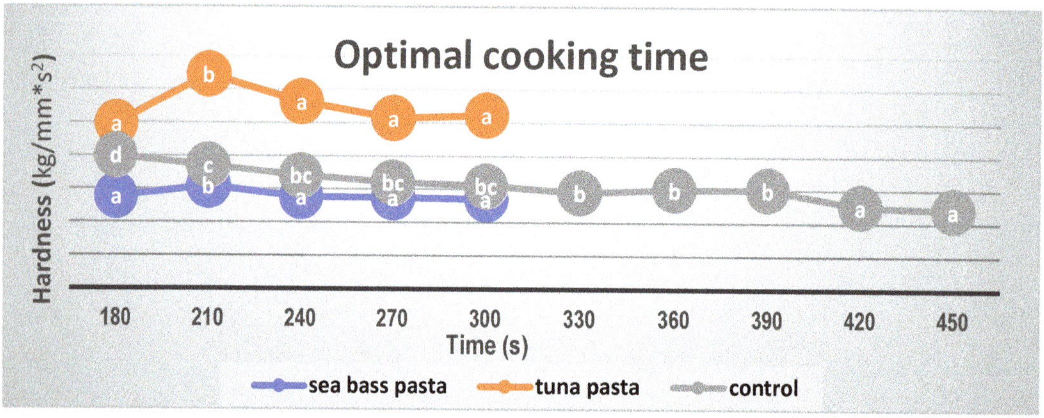

Figure 1. Optimal cooking time for tuna and sea bass pasta. Distinct letters indicate significant differences ($p < 0.05$) among the cooking times for each type of pasta.

3.3. *Texture Profile Analysis -TPA-*

As can be seen in Table 3, control pasta presented significantly higher hardness ($p < 0.05$) than enriched pasta with fish being significantly lower ($p < 0.05$) in tuna pasta. Related to fracturability, being higher in sea bass than tuna pasta. However, cohesiveness had a significantly higher value ($p < 0.05$) for tuna pasta which was similar to control pasta. On the other hand, gumminess and chewiness had a similar behavior being higher in control pasta than in pasta with fish. Related to adhesiveness, sea bass pasta seemed to have the same behavior as control pasta.

Table 3. TPA parameters for both types of enriched pasta and a control.

	HDN	ADH	SPG	COH	GUM	CHW	FRT
Control pasta	3726.35 ± 252.65 c	−16.02 ± 30.05 b	0.77 ± 0.09	0.68 ± 0.04 b	2545.55 ± 286.53 b	1993.26 ± 300.04 b	758.42 ± 86.63 b
Sea bass pasta	3206.25 ± 296.84 b	−13.79 ± 35.44 b	0.77 ± 0.09	0.49 ± 0.13 a	1610.34 ± 338.32 a	1279.20 ± 296.17 a	881.13 ± 125.81 c
Tuna pasta	2418.04 ± 304.11 a	−36.87 ± 20.34 a	0.72 ± 0.08	0.66 ± 0.06 b	1611.40 ± 291.77 a	1156.31 ± 239.39 a	479.18 ± 93.53 a

HDN: Hardness, ADH: Adhesiveness, SPG: springiness, COH: cohesiveness, GUM: Gumminess, CHW: chewiness, FRT: fracturability. Lowercase letters show significant differences among each type of pasta ($p < 0.05$).

For a better understanding of the relationship between physical and chemical variables, a PCA using a Pearson correlation matrix was developed. In essence, a multivariate analysis is a tool to simultaneously find patterns and relationships among several variables. It allows us to predict the effect that a change in one variable will have on the other variables [26]. In this regard, between TPA values and the fatty acid composition for each type of enriched pasta, a relationship could be detected for some parameters as could be seen in Figure 2.

Related to sea bass (Figure 2A), a correlation was shown for cohesiveness, gumminess and chewiness (F1). Cohesiveness showed a negative correlation with saturated, *trans* and polyunsaturated fatty acids while had a positive correlation (>0.89) with monounsaturated, especially oleic and linoleic showing up close at the PCA. Gumminess had a similar performance having a negative correlation with saturated and trans fatty acids. However, a positive correlation was found with polyunsaturated fatty acids showing on the right side of the PCA. Finally, chewiness had the same behavior as gumminess.

Regarding tuna pasta (Figure 2B), only chewiness had a significant correlation with some fatty acids. It was related to monounsaturated and polyunsaturated fatty acids whereas its behavior was contrary to saturated fatty acids. However, this parameter showed a negative correlation with polyunsaturated fatty acids (DPA -C22:5 Ω-3- and EPA) showing on the right side of the PCA.

3.4. Pasta Color

The color parameters are shown in Figure 3.

Related to luminosity (L*), control and sea bass pasta presented similar values which are higher than in tuna pasta. The opposite effect was observed in the red index (a*) being significantly higher ($p < 0.05$) in tuna pasta than in control and sea bass pasta, which shows similar values. In the case of the yellow index (b*), pasta with fish concentrate had the same behavior between them and control pasta had higher values than enriched fish pasta. Finally, the total color difference (ΔE) showed that it in tuna pasta was higher than sea bass pasta ($p < 0.05$), confirming a large effect over the color which depends on the species used.

3.5. Technological Properties

Values of technological parameters are shown in Table 4. The addition of fish in pasta formulation showed a significant decrease in weight gain (WG) with similar behavior in sea bass and tuna pasta. Related to the swelling index (SI), there were significant differences among all pasta assayed, the control pasta had the highest index and tuna pasta showed the lowest value. In the same way, sea bass and tuna pasta did not show differences in cooking losses (CL) while the control was significantly distinct from the rest. Finally, moisture (M) was higher ($p < 0.05$) in the control pasta while both kinds of enriched pasta had a similar value.

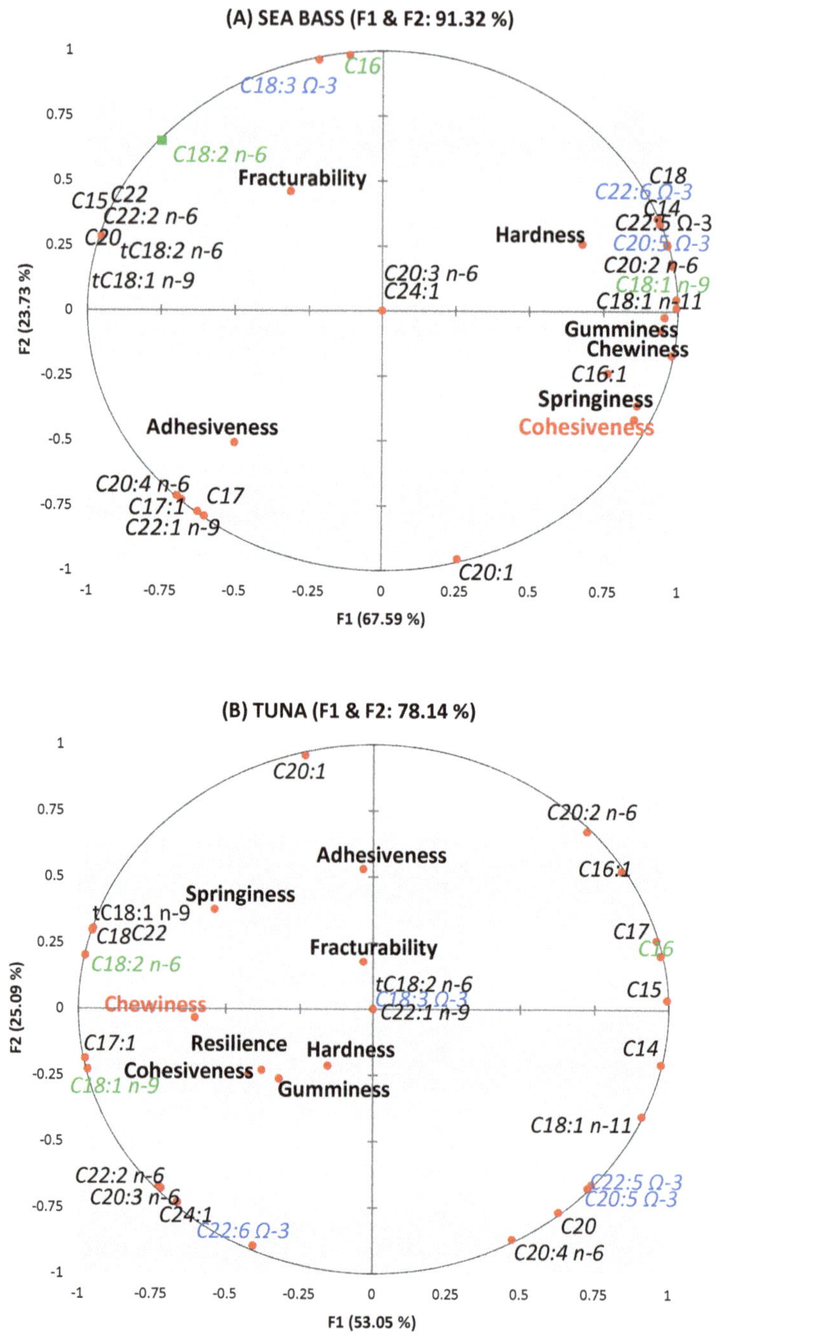

Figure 2. PCA of correlations between TPA parameters and fatty acids composition for sea bass pasta (**A**) and tuna pasta (**B**).

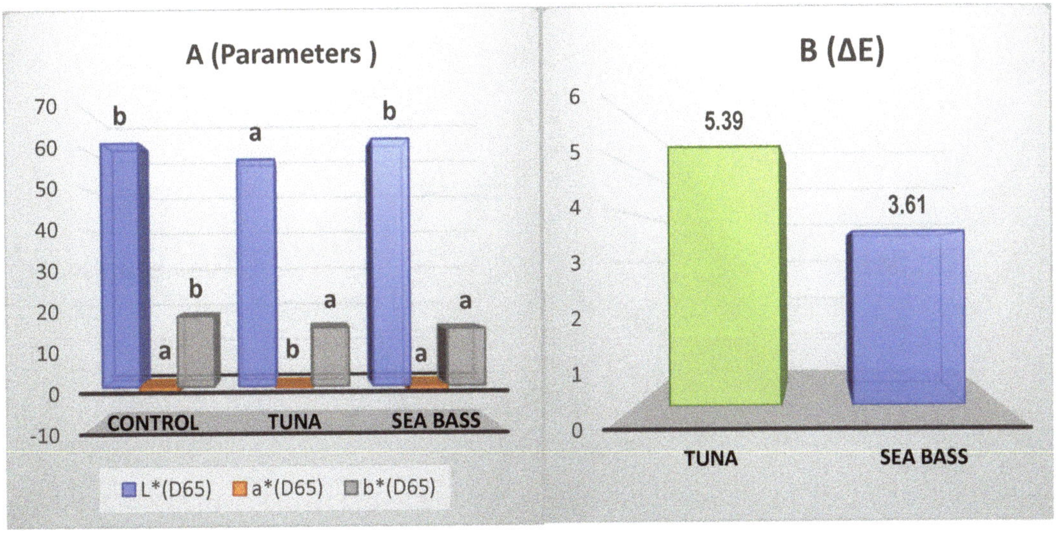

Figure 3. Color parameters for each assayed pasta (**A**) and variation with respect to control (**B**). Lowercase letters show significant differences among each type of pasta ($p < 0.05$).

Table 4. Values of technological properties for each enriched pasta developed.

	WG (%)	SI (g/g)	CL (%)	M (%)
Control	167.06 ± 3.15 b	3.20 ± 0.03 c	4.46 ± 0.17 a	11.59 ± 0.38 b
Sea bass	128.70 ± 8.80 a	2.72 ± 0.15 b	5.14 ± 0.75 b	10.58 ± 0.64 a
Tuna	129.35 ± 11.25 a	1.57 ± 0.04 a	4.91 ± 0.19 b	10.42 ± 0.21 a

WG: weight gain, SI: swelling index, CL: cooking losses, M: moisture. Lowercase letters show significant differences among types of pasta ($p < 0.05$)

3.6. Sensory Analysis

A radial graph (Figure 4) was made to compare the intensity of the studied attributes for each kind of enriched pasta compared with durum pasta (control).

According to ANOVA, significant differences were found for sensory attributes such as homogeneity, typical aroma, elasticity, fish flavor and fish odor. The control pasta had the highest value to homogeneity (9/10) while enriched pasta had a similar score (≈7/10). Typical aroma showed the same trend with a higher value for control pasta (8/10) than both enriched pasta. As in previous parameters, control pasta showed a higher value for elasticity (below 7). Regarding fish odor, as expected, the control pasta did not show this attribute whereas tuna had the highest value (6/10) followed by sea bass pasta (4/10). Related to the above, fish flavor was not present in the control pasta while sea bass had a lower value (≈1/10) than tuna pasta (5/10).

Sensory profiles of the pasta assessed are shown in Figure 5. Based on discriminatory power for each attribute assessed by the sensory panel only 8 of 11 were selected to draw profiles that turned out to be very different from each other. The first component (F1) collected 82.20% of the total variation and shows a clear separation between control pasta and enriched both kinds of pasta. Tuna was characterized by its fish odor ($p < 0.01$), fish flavor ($p < 0.01$) and after-taste. In contrast, the control pasta showed a typical aroma ($p < 0.05$) and it was found close to pasta characteristic flavor and elasticity ($p < 0.01$). However, in enriched pasta with sea bass no particular attribute stood out, being located in the plot between the control pasta and the tuna pasta. Due to their proximity, enriched pasta could resemble each other.

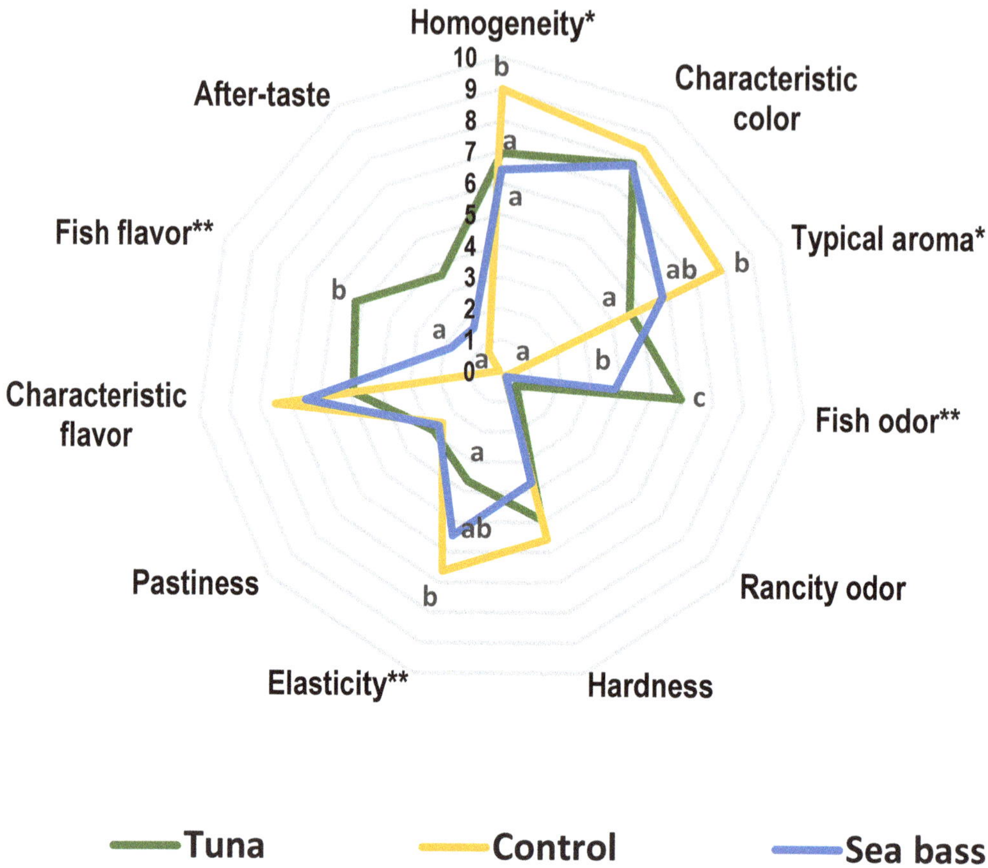

Figure 4. Radial graph of sensory attributes for enriched and control pasta. Lowercase letters show significant differences among the type of pasta. * Attribute showed significant differences ($p < 0.05$). ** Attribute showed high significant differences ($p < 0.01$).

In order to get a relation between sensory analysis and fatty acids composition in enriched pasta, a PCA was made and is shown in Figure 6.

The first component (F1) collected 48.08% of the variability and separated the different types of fish used. Tuna was related to fish odor and fish flavor and, with some polyunsaturated acid, in special those belonging C18 type, as well as C20:2 n-6 and C22:5 n-3. Additionally, rancidity odor, hardness and after-taste were associated with tuna too. The after-taste was highlighted due to its relationship with the *trans* C18:2 n-6. Fatty acids from cereals, (C18:2 n-6, C18:1 n-9 and C18:3 n-3) located on the left side of the plot and close to Tuna were related to rancidity odor. Conversely, on the other side of the plot, the sea bass was related to a typical aroma and characteristic flavor of the pasta. For its part, homogeneity, elasticity, and pastiness showed similar behavior and were located in the middle of the plot. The most outstanding attributes of the seabass pasta were associated with saturated fatty acids (C14, C15, C16, C17, C17, C20 and C22) and some polyunsaturated fatty acids as EPA (C22:6 n-3) and DHA (C20:5 n-3) which could be associated with the characteristic flavor of this kind of pasta.

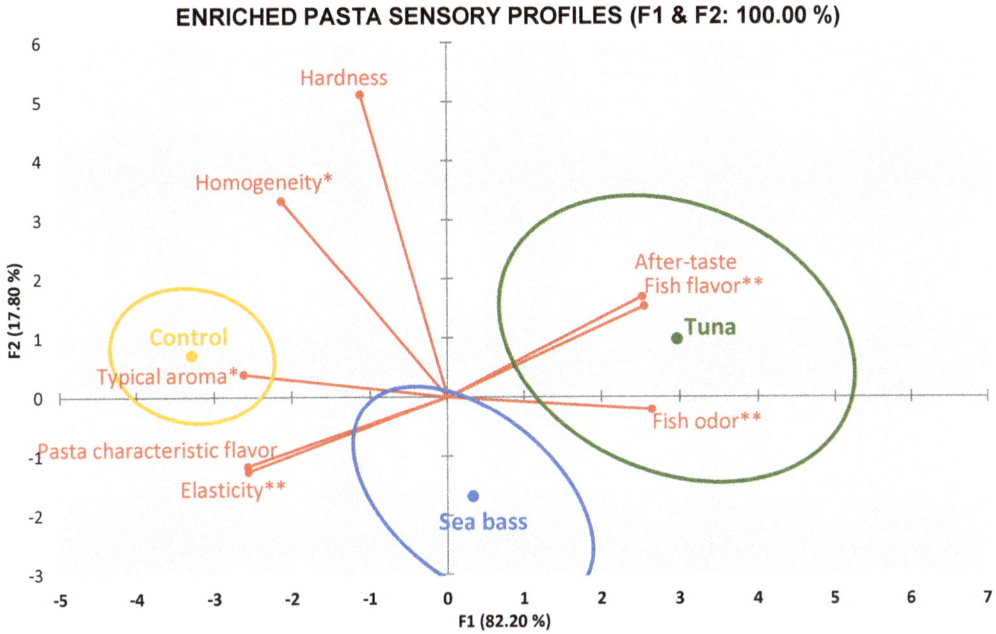

Figure 5. Sensory profile for fish enriched pasta and durum pasta (control). * Attribute showed significant difference ($p < 0.05$). ** Attribute showed significant difference ($p < 0.01$).

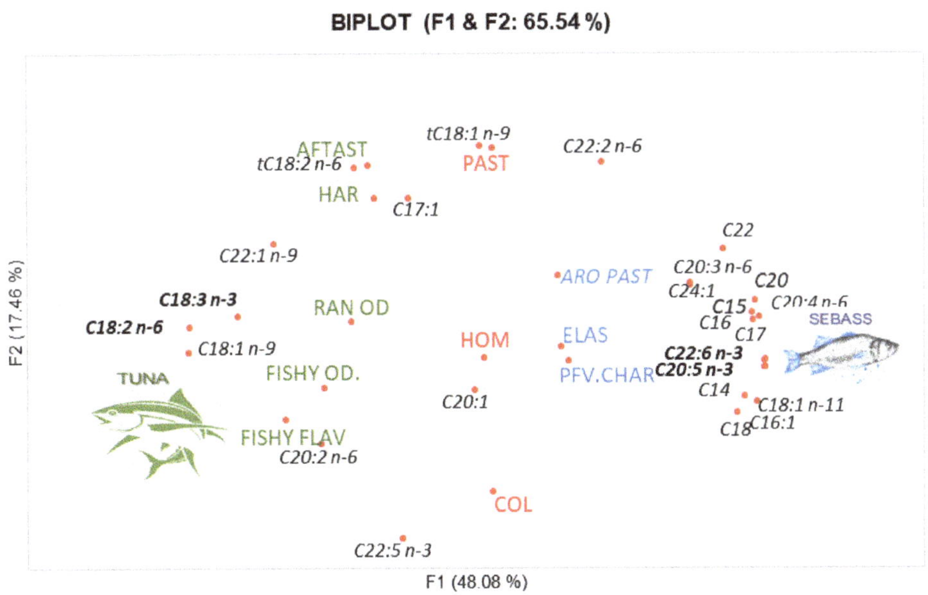

Figure 6. PCA from correlation matrix between sensory attributes and fatty acids composition for Sea bass pasta and Tuna pasta. AFTAST: after taste, HAR: hardness, RAN OD: rancidity odor, FISHY OD: Fishy odor, FISHY FLAV: Fishy flavor, PAST: Pastiness, ARO PAST: Pasta aroma, HOM: Homogeneity, COL: Typical color, ELAS: elasticity and PFV.CHAR. Pasta characteristic flavor.

4. Discussion

4.1. Fatty Acids Profile

As expected, pasta with tuna concentrate had a higher content of fatty acids than in sea bass concentrate, although it had some exceptions. Due to the last changes in the aquaculture feed, the increase in vegetable oils in the diet of Mediterranean farmed fish caused the oleic and linoleic content to be higher in pasta with sea bass than tuna, which resulted in an increase in %MUFA and content of Ω6 [24]. As seen in other studies, the content of EPA and DHA, the most important for a healthy diet [27], was higher in tuna, especially for the species used in this study (*Thunnus obesus*) [28,29]. Therefore, the behavior observed in these fish could be seen in pasta enriched with them. Concerning ratios, although the ideal ratio for Ω6/Ω3 was considered to be around 4:1 [30], this study found values higher in sea bass pasta. Some previous studies found ratios that agreed with our results [5,19]. On the other hand, EFSA made a recommendation of 300 mg for the Ω3 mg/100 g ratio for a day [31]. With the tuna pasta developed, we reached (80.97%) of dietary reference intake of Ω3, while with the sea bass pasta, it was reached 44.75% mg/100 g). Consequently, our findings demonstrated that enriched pasta with tuna or sea bass represented an adequate source to get the daily reference intake of PUFA (%DRI) reaching almost 81% for tuna pasta and 45% for sea bass pasta [32].

4.2. Cooking Times

As could be seen in the results, enriched pasta needed a lower time for cooking than control pasta. When fish is incorporated into pasta formulation, Physico-chemical characteristics are modified. The starch content decreases and for this reason, the water required for its gelatinization decreases too. The substitution of semolina implies a decrease in the glutenin content. Thus, less time to cook is required [33]. On top of that, the required time was higher for enriched pasta than in a previous study due to the percentage of fish in the formula [20].

4.3. Texture Profile Analysis

The effect of different added components such as starch, lipid or other ingredients could influence the pasta texture profile, especially for hardness [32]. In this way, the decrease in hardness parameter detected in enriched pasta was associated with the weakening of the structure due to the incorporation of lipids and proteins from fish meat that modify the matrix of gluten and starch [33]. Regarding fracturability, it is usually associated with hardness and was different between enriched pasta and control. It could be due to the behavior provided by myofibrillar proteins but especially by fat composition in each kind of pasta. According to other studies, texture properties could be modified with the incorporation of other ingredients different from semolina and resulted in unwanted textures [34]. Concerning cohesiveness, tuna and control pasta had the same values, whereas, in adhesiveness, sea bass and control were similar. These findings confirm the behavior of *trans* fatty acids like saturated fatty acids instead of other unsaturated in sea bass pasta. The above is a common characteristic of vegetables processed oils widely used in the animal feed industry. Conversely, tuna, a fish from the catch, had a higher value of adhesiveness which may be due to its unsaturated fatty acids quantities.

4.4. Color of Pasta

Brightness and red index (a*) had the same behavior. The incorporation of sea bass in pasta did not modify these parameters in comparison with enriched pasta with tuna due to the difference of color between these fish species. However, the yellow index (b*) was one of the most important parameters for pasta acceptability [35] and it was a characteristic color of pasta. For this reason, control pasta had the highest value for this index. The global color variation (ΔE) with respect to control pasta was higher in tuna pasta. Sea bass showed values in agreement with another study [9]. The results had variations in the range

3–6 as can be seen in Figure 3. Despite this, these changes cannot probably be seen with the naked eye [9].

4.5. Technological Parameters

The capacity of pasta to absorb water depends on its composition and processing conditions [36]. In this way, lipids and proteins from fish interact with starch for water absorption and reduce starch hydration, thus, enriched pasta had lower weight gain. On top of that, the reduced swelling index could be due to the formation of a protein network and different complexes between starch and lipids. These results were found in other studies [9,37]. On the other hand, the increase in cooking loss could be produced by the introduction of non-gluten proteins that weakened the network structure. Similar behavior had been shown in other studies of pasta fortified with other ingredients [4,9,38]. Finally, the values obtained for moisture were below those marked by regulation, which sets 12.5% of moisture for dry pasta [39].

4.6. Sensory Analysis

The addition of fish concentrates to enrich pasta caused some changes in their sensory profiles, especially providing odor and taste, as noted in previous studies [19,20]. This fact explained the increase in fish odor and flavor in enriched pasta being higher in tuna which presented more unsaturated fatty acids. Organoleptic properties obtained in this study are in agreement with those results reported by Devi (2013) [40] in pasta with incorporated fish. Furthermore, there were no remarkable changes in texture and appearance attributes due to the low percentage of fish concentrate added (3%). Regarding the above, earlier research that used tuna and tilapia mill meat to making laminated pasta (lasagnas) demonstrated there were no significant differences in these aspects either [41]. According to Figures 3 and 4, sea bass had an intermediate profile between control pasta and tuna pasta. It was similar to control in yellow color, pasta characteristic aroma and elasticity but differs from tuna pasta in the quantities of unsaturated fatty acids, especially Ω3 (EPA and DHA), being higher in tuna and therefore offering a typical smell and taste of fish. The above was corroborated in Figure 6 where fishy odor and flavor were related to fatty acids of tuna pasta.

In summary, it is possible to affirm that pasta could be enriched with both species of fish (tuna and seabass). However, the pasta profile depends on the type of fish used because each one provides completely organoleptic properties due to its composition in fatty acids as a consequence of its origin and diet.

5. Conclusions

The use of sea bass and tuna by-products to enrich pasta is an excellent alternative due to the contribution of protein and polyunsaturated fatty acids (Ω3) in these species. Tuna contributes three times as much DHA and EPA to pasta as sea bass. Therefore, the addition of tuna could be reduced from 3% to 1% in future studies to improve its sensory profile, which is characterized by a high fish flavor and fish odor compared to sea bass pasta and control pasta. Besides this, texture profile, color and technological quality parameters were modified by the addition of fish. The texture parameters showed a significant decrease in almost all parameters compared to control pasta as in the technological parameters, except for cooking losses, which were higher for enriched pasta than for control pasta. Finally, the sensory profiles of all pasta were adequate, showing a better behavior in sea bass pasta in comparison with control pasta.

Author Contributions: Conceptualization, A.A., J.B.C.M. and J.A.B.; methodology, A.A., A.H. and P.L.M.; software, A.A. and J.B.C.M.; validation, A.A., A.H. and P.L.M.; formal analysis, J.A.B., P.R. and J.B.C.M.; investigation, A.A., A.H., J.A.B. and J.B.C.M.; resources, J.A.B.; data curation, A.A. and J.B.C.M.; writing—original draft preparation, A.A., A.H. and J.B.C.M.; writing—review and editing, A.A.; visualization, J.B.C.M. and P.R.; supervision, J.A.B. and J.B.C.M.; project administration, J.A.B. and J.B.C.M.; funding acquisition, J.A.B. All authors have read and agreed to the published version of the manuscript.

Funding: This research received no external funding.

Institutional Review Board Statement: Not applicable.

Informed Consent Statement: Not applicable.

Conflicts of Interest: The authors declare no conflict of interest.

References

1. Aronson, J.K. Defining 'nutraceuticals': Neither nutritious nor pharmaceutical. *Br. J. Clin. Pharmacol.* **2016**, *83*, 8–19. [CrossRef]
2. Khan, R.S.; Grigor, J.; Winger, R.; Win, A. Functional food product development—Opportunities and challenges for food manufacturers. *Trends Food Sci. Technol.* **2013**, *30*, 27–37. [CrossRef]
3. Siró, I.; Kápolna, E.; Kápolna, B.; Lugasi, A. Functional food. Product development, marketing and consumer acceptance—A review. *Appetite* **2008**, *51*, 456–467. [CrossRef]
4. Kadam, S.; Prabhasankar, P. Marine foods as functional ingredients in bakery and pasta products. *Food Res. Int.* **2010**, *43*, 1975–1980. [CrossRef]
5. Iafelice, G.; Caboni, M.F.; Cubadda, R.; Di Criscio, T.; Trivisonno, M.C.; Marconi, E. Development of Functional Spaghetti Enriched with Long Chain Omega-3 Fatty Acids. *Cereal Chem. J.* **2008**, *85*, 146–151. [CrossRef]
6. Jacobsen, C. Enrichment of foods with omega-3 fatty acids: A multidisciplinary challenge. *Ann. N. Y. Acad. Sci. USA* **2010**, *1190*, 141–150. [CrossRef]
7. Filipović, J.; Pezo, L.; Filipovic, V.; Brkljača, J.; Krulj, J. The effects of ω-3 fatty acids and inulin addition to spelt pasta quality. *LWT* **2015**, *63*, 43–51. [CrossRef]
8. Reglamento (UE) No 116/2010 de la Comisión, de 9 de Febrero de 2010, por el que se Modifica el Reglamento (CE) no 1924/2006 del Parlamento Europeo y del Consejo en lo Relativo a la Lista de Declaraciones Nutricionales Texto Pertinente a Efectos del EEE. Available online: https://www.boe.es/buscar/doc.php?id=DOUE-L-2010-80235 (accessed on 13 May 2021).
9. Desai, A.; Brennan, M.A.; Brennan, C.S. The effect of semolina replacement with protein powder from fish (Pseudophycis bachus) on the physicochemical characteristics of pasta. *LWT* **2017**, *89*, 52–57. [CrossRef]
10. Goes, E.S.D.R.; De Souza, M.L.R.; Michka, J.M.G.; Kimura, K.S.; De Lara, J.A.F.; Delbem, A.C.B.; Gasparino, E. Fresh pasta enrichment with protein concentrate of tilapia: Nutritional and sensory characteristics. *Food Sci. Technol.* **2016**, *36*, 76–82. [CrossRef]
11. Munekata, P.E.S.; Pateiro, M.; Domínguez, R.; Zhou, J.; Barba, F.J.; Lorenzo, J.M. Nutritional Characterization of Sea Bass Processing By-Products. *Biomolecules* **2020**, *10*, 232. [CrossRef]
12. Hepburn, F.N.; Exler, J.; Weihrauch, J.L. Provisional tables on the content of omega-3 fatty acids and other fat components of selected foods. *J. Am. Diet. Assoc.* **1986**, *86*, 788–793. [PubMed]
13. Medina, I.; Aubourg, S.P.; Martín, R.P. Composition of phospholipids of white muscle of six tuna species. *Lipids* **1995**, *30*, 1127–1135. [CrossRef]
14. Sharma, S.K.; Bansal, S.; Mangal, M.; Dixit, A.K.; Gupta, R.K.; Mangal, A. Utilization of Food Processing By-products as Dietary, Functional, and Novel Fiber: A Review. *Crit. Rev. Food Sci. Nutr.* **2015**, *56*, 1647–1661. [CrossRef]
15. Jayathilakan, K.; Sultana, K.; Radhakrishna, K.; Bawa, A.S. Utilization of byproducts and waste materials from meat, poultry and fish processing industries: A review. *J. Food Sci. Technol.* **2011**, *49*, 278–293. [CrossRef] [PubMed]
16. Sibbel, A. The sustainability of functional foods. *Soc. Sci. Med.* **2007**, *64*, 554–561. [CrossRef]
17. European Commission. Report from the Commission to the European Parliament, the Council, the European Economic and Social Commitee and the Commitee of the Regions on the Implementation of the Circular Economy Action Plan. 2019. EUR-Lex-52019DC0190. Available online: https://eur-lex.europa.eu/legal-content/EN/TXT/?uri=CELEX%3A52019DC0190 (accessed on 15 June 2021).
18. United Nations. The 2030 Agenda for Sustainable Development. Available online: https://sdgs.un.org/es/goals (accessed on 15 June 2021).
19. Calanche, J.; Beltrán, H.; Marquina, P.; Roncalés, P.; Beltrán, J. Eating fish in another way: Development of functional pasta with added concentrates of farmed sea bass (Dicentrarchus labrax). *Cereal Chem. J.* **2019**, *96*, 856–865. [CrossRef]
20. Ainsa, A.; Marquina, P.; Roncalés, P.; Calanche M, J.B. Enriched Fresh Pasta with a Sea Bass By-Product, a Novel Food: Fatty Acid Stability and Sensory Properties throughout Shelf Life. *Foods* **2021**, *10*, 255. [CrossRef] [PubMed]
21. Cleary, L.J.; Brennan, C. The influence of a (1?3)(1?4)-?-d-glucan rich fraction from barley on the physico-chemical properties and in vitro reducing sugars release of durum wheat pasta. *Int. J. Food Sci. Technol.* **2006**, *41*, 910–918. [CrossRef]

22. AACC. *Approved Methods of the American Association of Cereal Chemists*, 10th ed.; AACC: St. Paul, MN, USA, 2000.
23. Larson-Powers, N.; Pangborn, R.M. Descriptive analysis of the sensory properties of beverages and gelatins containing sucrose or synthetic sweeteners. *J. Food Sci.* **1978**, *43*, 47–51. [CrossRef]
24. UNE-EN ISO (2012). *Análisis Sensorial. Guía General Para La Selección, Entrenamiento y Control de Catadores y Catadores Expertos*; ISO: Geneva, Switerland, 2013.
25. Braghieri, A.; Piazzolla, N.; Carlucci, A.; Monteleone, E.; Girolami, A.; Napolitano, F. Development and validation of a quantitative frame of reference for meat sensory evaluation. *Food Qual. Prefer.* **2012**, *25*, 63–68. [CrossRef]
26. Calanche, J.; Pedrós, S.; Roncalés, P.; Beltrán, J.A. Design of Predictive Tools to Estimate Freshness Index in Farmed Sea Bream (Sparus aurata) Stored in Ice. *Foods* **2020**, *9*, 69. [CrossRef]
27. Babuskin, S.; Krishnan, K.R.; Saravana, P.A.; Sivarajan, M.; Sukumar, M. Functional Foods Enriched with Marine Microalga Nannochloropsis Oculata as a Source of W-3 Fatty Acids. *Food Technol. Biotechnol.* **2014**, *52*, 292–299.
28. Zhang, J.; Tao, N.; Zhao, Y.; Wang, X.; Wang, M. Comparison of the Fatty Acid and Triglyceride Profiles of Big Eye Tuna (Thunnus obesus), Atlantic salmon (Salmo salar) and Bighead Carp (Aristichthysnobilis) Heads. *Molecules* **2019**, *24*, 3983. [CrossRef]
29. Peng, S.; Chen, C.; Shi, Z.; Wang, L. Amino Acid and Fatty Acid Composition of the Muscle Tissue of Yellowfin Tuna (Thunnus Albacares) and Bigeye Tuna (Thunnus Obesus). *J. Food Nutr. Res.* **2013**, *4*, 42–45.
30. Candela, C.G.; López, L.M.B.; Kohen, V.L. Importance of a balanced omega 6/omega 3 ratio for the maintenance of health: Nutritional recommendations. *Nutr. Hosp.* **2011**, *26*. [CrossRef]
31. ISSFAL (International Society for the Study of Faty Acids and Lipids). Global Recommendations for EPA and DHA Intake. 2014. Available online: https://www.issfal.org/ (accessed on 25 May 2021).
32. Products, N.E.P.O.D. Scientific Opinion on Dietary Reference Values for fats, including saturated fatty acids, polyunsaturated fatty acids, monounsaturated fatty acids, trans fatty acids, and cholesterol. *EFSA J.* **2010**, *8*. [CrossRef]
33. Vernaza, M.G.; Biasutti, E.; Schmiele, M.; Jaekel, L.Z.; Bannwart, A.; Chang, Y.K. Effect of supplementation of wheat flour with resistant starch and monoglycerides in pasta dried at high temperatures. *Int. J. Food Sci. Technol.* **2012**, *47*, 1302–1312. [CrossRef]
34. De Marco, E.R.; Steffolani, M.E.; Martínez, C.S.; León, A.E. Effects of spirulina biomass on the technological and nutritional quality of bread wheat pasta. *LWT* **2014**, *58*, 102–108. [CrossRef]
35. Pongpichaiudom, A.; Songsermpong, S. Evaluation of microstructure and quality characteristics of microwave-dried instant noodles enriched with chicken meat, egg yolk, and seaweed. *J. Food Meas. Charact.* **2017**, *12*, 22–34. [CrossRef]
36. Marti, A.; Fongaro, L.; Rossi, M.; Lucisano, M.; Pagani, M.A. Quality characteristics of dried pasta enriched with buckwheat flour. *Int. J. Food Sci. Technol.* **2011**, *46*, 2393–2400. [CrossRef]
37. Liu, T.; Hamid, N.; Kantono, K.; Pereira, L.; Farouk, M.; Knowles, S.O. Effects of meat addition on pasta structure, nutrition and in vitro digestibility. *Food Chem.* **2016**, *213*, 108–114. [CrossRef] [PubMed]
38. Lacko-Bartošová, M. Noodle quality of winter wheat cultivated in sustainable farming systems. *J. Central Eur. Agric.* **2014**, *15*, 84–94. [CrossRef]
39. Real Decreto 2181/1975, de 12 de Septiembre, por el que se Aprueba la Reglamentación Técnico-Sanitaria para la Elaboración, Circulación y Comercio de Pastas Alimenticias. Boletín Oficial del Estado No 220, de Septiembre de 1975. Available online: https://www.boe.es/eli/es/d/1975/09/12/2181 (accessed on 4 May 2021).
40. Devi, L. Utilization of Fish Mince in Formulation and Development of Pasta Products. *Int. Food Res. J.* **2013**, *20*, 219–224.
41. Kimura, K.S.; De Souza, M.L.R.; Gasparino, E.; Mikcha, J.M.G.; Chambó, A.P.S.; Verdi, R.; Coradini, M.F.; Marques, D.R.; Feihrmann, A.; Goes, E.S.D.R. Preparation of lasagnas with dried mix of tuna and tilapia. *Food Sci. Technol.* **2017**, *37*, 507–514. [CrossRef]

foods

Article

Evaluation of Glycemic Index of Six Different Samples of Commercial and Experimental Pasta Differing in Wheat Varieties and Production Processes

Amalia Pandolfo, Bernardo Messina and Giuseppe Russo *

Gian Pietro Ballatore Research Consortium, Z.I. Dittaino I, 94010 Assoro, EN, Italy; pandolfo.amalia@gmail.com (A.P.); info@ilgranoduro.it (B.M.)
* Correspondence: giusepperusso@ilgranoduro.it

Abstract: Pasta is a staple food of the Mediterranean Diet, and it is traditionally made of durum wheat semolina. In Sicily, durum wheat production and its transformation into semolina, bread, and pasta are well-developed economic sectors. For pasta, there is a wide supply of commercial brands, whether coming from conventional industrial manufacturing or from medium to small and local handcrafted production. Both conventional durum wheat and local durum wheat landraces, such as *Timilia* and *Russello*, are used for pasta production, but local landraces are, for the most, transformed into handcrafted pasta. The market of local landraces durum wheat pasta has risen in recent decades, in Sicily and in Italy as well, boosted by a perceived high nutritional and healthy value of these wheat derivatives. In particular, a popular and scientifically unproven idea suggests that a reduced glycemic response might be elicited by these pasta landraces. Therefore, to test this hypothesis, the main objective of the present study was the evaluation of the glycemic index (GI) of four samples of *Timilia* and *Russello* handcrafted pasta and two samples of conventional durum wheat pasta. The study enrolled fourteen healthy weight male and female volunteers aged from 18 to 46; eight test sessions were performed twice a week, every session testing a pasta sample (six sessions) or the glucose solution chosen as reference food (two sessions). The standard methodology for GI measurement was followed during each step of the study. The six tested pasta samples were characterized regarding their composition (protein, fiber, and starch content) and their whole production processes (milling method and milling diagram of flour or semolina, drying temperature, and diagram of pasta shape). The six tested pasta samples showed GI values ranging from low (34.1) to intermediate (63.1). *Timilia* and *Russello* pasta are the first GI calculations available. The two samples made of conventional grains showed lower values of GI (34.1 and 37.8). The results do not support the popular idea of a reduced glycemic response elicited by *Timilia* and *Russello* wheat landrace pasta; the tested samples showed GI values in the range of 56.2 to 63.1. However, some consideration should be made of factors other than wheat varieties and related to production processes that may have affected the final GIs of the pastas. Even if the study is not designed to discriminate among factors related to wheat varieties or processes used to produce different pasta, it is a preliminary step in the characterization of the healthy potential of the local wheat landraces, popularly called ancient grain. A future implementation of the local wheat landraces supply chain should pay attention to all the factors above, from a better seed identity certification to the production process in order to further improve the healthy value of these staples of the Mediterranean Diet.

Keywords: pasta; glycemic index; glycemic response; durum wheat; landraces

Citation: Pandolfo, A.; Messina, B.; Russo, G. Evaluation of Glycemic Index of Six Different Samples of Commercial and Experimental Pasta Differing in Wheat Varieties and Production Processes. *Foods* **2021**, *10*, 2221. https://doi.org/10.3390/foods10092221

Academic Editor: Claudia Monika Haros

Received: 18 June 2021
Accepted: 6 August 2021
Published: 18 September 2021

1. Introduction

The glycemic index (GI) was introduced in 1981 [1] as a way of classifying carbohydrate-rich foods according to their effects on postprandial glycemia.

In detail, the GI measures the postprandial blood glucose response of a 50 g portion of available carbohydrate as a percentage of the blood glucose response elicited by 50 g of a reference carbohydrate, such as glucose or white bread [1,2].

For both test and reference foods, the Incremental Area Under the Curve (IAUC) of blood glucose concentration in two hours is calculated. The ratio percentage of the two areas is the GI for the tested food [2–4].

For a starchy food, there are many factors able to more or less make the enzymatic release of its glucose content in the intestinal lumen and/or its absorption easy, leading to a major or minor increase in blood glycemia of a subject, regardless of their physiological background.

For pasta, the focus food in this study, durum wheat varieties, pasta composition (fiber, protein, and starch content), structure and dimensions of the raw starch granules, milling processes and flour or semolina grinding diagram, drying temperature and diagram, pasta shape, starch gelatinization degree after cooking and cooking time [5–8] are all factors affecting the glycemic response after pasta ingestion, and as a consequence, its final GI [9,10].

Since its introduction, several papers have been published pointing out a method for determining GI of different foods [2–4] and assessing the health impact of diets based on low, moderate, or high GI foods on the outcome, such as metabolic syndrome, diabetes prevention and control, prevention of cardiovascular disease (CVD), reduction and prevention of obesity, and prevention of some cancers, i.e., breast and colon cancers [5,9–11]. One point for wide and useful use of GI and GL in dietary recommendations is the availability of reliable GI data for the food items actually consumed by people in a specific country.

Most of the existing tables report the calculated GI for foods usually available and eaten in the United States or other Western countries [1,12,13], whilst less GI data has been collected and reported for foods usually manufactured or eaten in the Mediterranean countries and in Italy.

A recent contribution that has partially filled the gap is the paper of Scazzina et al. (2016), reporting the GI of 124 different food items easily available in Italian markets [14].

The current study evaluated for the first time the GI of four pasta samples, produced from durum wheat landraces cultivated in Sicily (*Russello* and *Timilia*), using the standard methodology for GI measurement.

In addition, two other samples obtained from industrial pasta factories, one commercial wholemeal pasta and one experimentally produced pasta (not wholemeal), were added to the study.

Timilia and *Russello* Sicilian wheat landraces, locally named "ancient grain," were chosen because of the recent sharp increase in their local customer demand, related to a hypothetical higher healthy value of their derivatives [15–21].

Namely, it is generally assumed that derivatives of these traditional landraces might elicit a reduced glycemic response, but, thus far, there is no scientific data supporting this specific view.

The study also collected data on tested pasta composition and on the processes for their production, as well as an attempt to draw attention to variables that may have affected the final GI values measured in the study.

Therefore, the study should be considered a contribution to a better characterization of local wheat landrace derivatives.

Moreover, it points to classify the six pasta samples analyzed according to Ramdath [5], which defines food containing starch or carbohydrates as "High" (GI range from 100 to 70), "Intermediate" (GI range from 69 to 65), and "Low" (GI less than 55) glycemic index classes.

2. Materials and Methods

2.1. Study Design

The glycemic indexes (GIs) of 4 Timilia and Russello handcrafted pasta samples and 2 samples of conventional durum wheat pasta were evaluated. The GI was determined according to Wolever (2010), Brouns et al. (2005), and the FAO/WHO (1998) method and procedures [2–4]. A total of 14 healthy weight male and female volunteers, aged from 18 to 46 were enrolled in the study, and 8 test sessions were performed twice a week, every session testing a pasta sample (6 sessions) or the glucose solution chosen as reference food (2 sessions).

2.2. Logistic and Protocol Requirements

The hotel management school, *Istituto Professionale di Stato per i Servizi Alberghieri e della Ristorazione "Paolo Borsellino"* (Palermo, Italy), made available suitably equipped rooms and a professional kitchen for the study; the kitchen staff was trained to cook and serve each sample according to the study design.

A team of 2 researchers supervised the tests and collected field data. In total, 8 evaluation tests were scheduled, 1 for each pasta sample and 2 additional sessions for reference food (50 g of glucose in 250 mL solution). On each test day, a single sample, pasta or reference food, was tested on every volunteer.

The volunteers fasted overnight and were instructed to keep their usual dietetic habit and to repeat the same standardized dinner by 7:00–8:00 p.m., each evening before the tests. No alcohol or intense physical activity was allowed the day before the test. The test started at 7:00 a.m. and ended at 9.30 a.m.

During the 6 pasta testing sessions, subjects were asked to drink 250 mL of mineral water provided with the serving (labeled "Acqua Vera"). They were asked to consume the sample and drink the water in not more than 10 min, and to stay sitting for the next 2 h without eating or drinking anything.

Three specialized nurses took care of sampling fingertip capillary blood at 0 (fasted state) and at 15, 30, 45, 60, 90, 120 min from food ingestion. Glucometers (ACCU-CHEK-AVIVA; labeled Roche Diagnostics GmbH, Mannheim, Germany) were used to measure blood glucose concentration (mmol). The choice of the glucometer device was settled by suggestions of Freackman et al. 2012 [22] that argued the best type of systems available for guaranteeing accuracy evaluation according to DIN EN ISO 15.197: 2003.

The measured glycemia values were used to build the curve of glycemic response for every volunteer and for each tested and reference food.

2.3. Healthy Volunteers

Totally, 29 subjects from the general population responded to a public claim for trial recruiting. Before recruiting, each one of the subjects had an interview with a physician and an accredited dietitian to verify their good health and suitability to the purpose of the study.

For each of them, weight, height, and body mass index (BMI) were measured and calculated. Volunteers were individually instructed on the aim of the study, on the protocol that would be applied, and informed about the risks of the procedure. Of the 17 initially recruited subjects, only one was labeled as IG15 withdrawn, while the other two subjects (IG09 and IG14) were discarded from the study, having missed some of the test sessions. Therefore, final data were available for 14 subjects who took part in the 8 scheduled test sessions from 10 November 2015 to 4 December 2015.

The sessions were performed twice a week, allowing at least 48 h of wash-out between 2 following test days. Only 14 volunteers completed every session of the study. The volunteers were adult males (n = 14) and females (n = 3) aged from 18 to 48 years. A label from IG01 to IG17 was assigned (Table 1). As reported and explained in the results, just 3 of the volunteers withdrew from the study.

Table 1. Data on volunteers enrolled in the study (BMI = Body Mass Index).

Label	Gender	Age	Weight (kg)	Height (cm)	BMI
IG01	M	18	69.5	166	25.0
IG02	M	19	78	178	25.0
IG03	M	32	73	175	23.8
IG04	F	26	64	170	22.1
IG05	F	46	67	167	24.0
IG06	M	26	98	195	25.8
IG07	M	26	80	177	25.5
IG08	M	19	91.5	188	25.5
IG10	M	25	74	170	25.6
IG11	M	25	72	175	23.5
IG12	M	45	71	174	23.5
IG13	F	43	54	158	21.6
IG16	M	40	85	181	25.9
IG17	M	38	94	193	26.0
Mean		30.6	76.2	176.2	24.5
SD		10.0	12.7	10.4	1.4

2.4. Tested Food

The trial was designed to measure the GIs of 6 different types of pasta, whose characteristics are reported in detail in Table 2. The 6 samples were labeled as follows:

- ICS experimental pasta (shape spaghetti) produced by the industrial plant using a mix of varieties of durum wheat cultivated in Sicily (*Duilio, Simeto, Iride*, and *Saragolla*).
- COMM commercial wholewheat pasta (shape spaghetti) produced by means in an industrial plant using an unknown mix of varieties of durum wheat.
- TI-DIMESA experimental handcrafted pasta (shape tagliatelle) from *Timilia* Sicilian landrace.
- TI-PONTE commercial handcrafted pasta (shape tagliatelle) from *Timilia* Sicilian landrace.
- RU-DIMESA experimental handcrafted pasta (shape tagliatelle) from *Russello* Sicilian landrace.
- RU-PONTE commercial handcrafted pasta (shape tagliatelle) from *Russello* Sicilian landrace.

Table 2. Characteristics of samples of pasta analyzed.

Pasta Sample	Varieties or Landraces	Whole Wheat Pasta	Milling Process	Note on Flour	Drying Process	Protein Content (%)	TDF (%DM)	RS (%DM)
ICS	Mix of modern varieties *Duilio, Simeto, Saragolla, Iride*	NO	Cylinder grinding	Semolina totally deprived of "farinette" fraction [1]	Dynamic, at high temperature (90°)	12.50	4.65	0.76
COMM	Mix of unknown modern varieties (commercial pasta)	YES	Unknown	Whole Semolina	Unknown	12.00	6.50	0.47
RU-DIMESA	*Russello* Sicilian landrace	YES	Stone milling	More than 20% of "farinette" [1]	Static, at temperature around 30 °C	11.24	8.43	0.68
RU-PONTE	*Russello* Sicilian landrace	YES	Stone milling	More than 20%of "farinette" [1]	Static, at temperature under 40 °C	10.50	9.68	0.23
TI-DIMESA	*Timilia* Sicilian landrace	YES	Stone milling	More than 20% of "farinette" [1]	Static, at temperature around 30 °C	10.83	8.13	0.48
TI-PONTE	*Timilia* Sicilian landrace	YES	Stone milling	More than 20% of "farinette" [1]	Static, at temperature under 40 °C	10.60	9.61	0.51

[1] We consider as "farinette" the fraction of flour with a particle diameter $\leq$118 µ. TDF = Total Diet Fiber, RS = Resistant Starch, DM = Dry Matter.

As reference food, a solution containing 50 g of glucose in mineral water for 250 milliliters of volume (water labeled as "Acqua Vera") was chosen. The reference food was prepared and bottled by accredited chemists (Farmacia Sanfilippo-Palermo).

The reference food was tested twice, while each sample of pasta was tested once for each volunteer. For each type of pasta previously calculated, the sample serving size contained 50 g of available carbohydrates, and cooking time was tested (Table 3). Indeed, cooking time is well known as an important parameter affecting the pasta GI, and is often reported in official GI tables of food.

Table 3. Scheme to evaluate 50 g of available carbohydrates for each sample serving.

Pasta Samples		Cooking Time Minutes		Total Starch % Dry Matter	Total Starch %	Serving Containing 50 g Available CHO
ICS	raw		raw	76.39	70.81	70.60
	cooked	13	cooked	78.23	71.43	70.00
COMM	raw		raw	72.63	64.03	78.08
	cooked	11	cooked	73.69	67.32	74.27
RU-DIMESA	raw		raw	70.28	63.52	78.71
	cooked	8	cooked	70.08	64.43	77.61
RU-PONTE	raw		raw	69.89	61.15	81.75
	cooked	9	cooked	66.57	39.15	127.71
TI-DIMESA	raw		raw	67.43	60.21	83.03
	cooked	8	cooked	57.77	52.40	95.43
TI-PONTE	raw		raw	74.58	65.22	76.66
	cooked	9	cooked	70.37	40.90	122.25

The kitchen staff and one of the researchers had previously created the needed tests to calculate the optimal cooking time for each pasta sample.

Unsalted boiling water was chosen to cook the pasta to the preferred cooking point in Italy named "cottura al dente," corresponding to the time when samples cooked were firm to the bite and not sticky during chewing [23].

At this point, the portions of pasta were quickly served without any sauce.

Pasta ICS is an experimental product shaped as spaghetti and made by a mix of semolina from four "modern" varieties of Sicilian durum wheat, namely *Duilio*, *Iride*, *Saragolla*, and *Simeto*.

Pasta ICS production took place at the factory *Molino e Pastificio Tomasello*, in Trabia (Palermo, Italy). The semolina used to produce pasta ICS showed a specific grinding diagram characterized by the total removal by extraction of the so-called "farinette" fraction, made up of particles with a diameter ≤118 μ, namely the less sized fraction obtained after milling durum wheat (according to UNI 10873:2000, which is the Italian standard to determine and classify durum wheat Semolina granulometry) [24]. As the last step of its production process, pasta ICS underwent a dynamic drying process, which used a temperature diagram always over 80–90 °C.

Pasta COMM is a wholegrain commercial pasta (spaghetti) available in a well-known Italian large-scale retail distribution. The shape was spaghetti, and the sample was produced with the pasta factory industrial method, not handcrafted. No other information about durum wheat varieties or milling diagrams was available for this sample.

TI-DIMESA e TI-PONTE are 2 wholegrain samples of pasta, shaped as tagliatelle and obtained from an artisanal production process. The flour was obtained by stone milling durum wheat of landrace *Timilia*. The flour was analyzed in the laboratory to evaluate the particle size profile using a Buhler automatic sieve. The analysis results showed that more than 20% of the flour particles had a diameter ≤118 μ. The 2 *Timilia* pasta samples underwent a drying process with a low-temperature drying diagram (30 °C or 40 °C) with a static drying cell.

In addition, RU-DIMESA e RU-PONTE are 2 wholegrain samples of pasta, shaped as tagliatelle, made up from a stone-milled flour of durum wheat landrace, *Russello*. This flour was analyzed in the laboratory to evaluate the particle size profile by Buhler automatic sieve confirming the presence of particles with a diameter ≤118 μ at more than 20% of the total. The *Russello* tagliatelle pasta was obtained from an artisanal production process and dried by a low-temperature drying diagram (30 °C or 40 °C) with a static drying cell.

Official methods were used to analyze every pasta sample to determine protein content (Kjeldhal AACC 47-12), moisture (UNI EN ISO 712:2010), total dietary fiber (Megazyme kit based on AACC method 32-05.01 and AOAC method 985.29), and resistant and total starch content (Megazyme kit K-TSTAR based on AACC 76.13).

2.5. Data Collection and Statistical Analysis

The measured glycemic values were used to build the curve of glycemic response for every volunteer and for every tested food, including the reference one. Then, for each sample and each study subject, the incremental area under the blood glucose response curve (IAUC) was calculated geometrically, using the trapezoid rule, and ignoring the area below the fasting baseline. The GI calculation for each pasta sample used the method referred to as the mean of the ratios. For each subject, the ratio between the individual IAUC after consuming the pasta sample and the IAUC for the same subject after consuming the reference food was calculated, expressed as a percentage value. Then, the GI of each pasta type was calculated as the average value of the ratios across all the subjects consuming the pasta sample [2–4]. One-way analyses of variance (ANOVA) were used to compare differences between the eight evaluation sessions with post-hoc Tukey HSD test. A p-value ≤ 0.05 was considered statistically significant. Pairwise comparison of glycemic index among all pasta samples ANOVA was also carried out.

3. Results

The results revealed, for the six tested samples, GI values ranging from 34.1 to 63.1 (Table 4). The lowest GI (34.1) was obtained for the pasta ICS, followed by pasta COMM, which showed a GI value of 37.8. According to the current classification [5], both these pastas should be considered low GI starchy foods. Pasta COMM is a wholegrain sample likely obtained from an industrial process as the ICS sample, but we have no details about durum wheat varieties used for its production. The other four pastas made with landraces Timilia (TI-DIMESA and TI-PONTE) or Russello (RU-DIMESA and RU-PONTE) showed higher GI values ranging from 56.2 to 63.1, making these pasta assessed as intermediate GI foods.

Table 4. Glycemic Index evaluated for the six pasta samples.

Samples	GI	SD
ICS	34.1	15.2
COMM	37.8	21.8
RUDIMESA	59.2	21.2
RUPONTE	56.2	18.0
TIDIMESA	57.1	27.4
TIPONTE	63.1	18.6

SD = Standard Deviation.

Three of the samples, in the order of RU-PONTE, TI-DIMESA and RU-DIMESA, showed close values (56.2, 57.1, and 59.2) and just the sample TI-PONTE (Timilia) registered the higher value of 63.1, which was also the highest calculated in the study and the higher value if compared with recent data recorded for Italian pasta [14].

Post-hoc analysis for pairwise comparison (Table 5) showed statistically significant higher GI values of TI-PONTE compared to ICS (difference = 29.0; 95% C.I. 6.1~51.9; p-value = 0.005) and COMM (difference = 25.3; 95% C.I. 2.0~48.6; p-value = 0.026), higher GI values of RU-DIMESA compared to ICS (difference = 25.1; 95% C.I. 2.2~48.0; p-value = 0.024) and TI-DIMESA compared to ICS (difference = 23.0; 95% C.I. 0.1~45.9; p-value = 0.049). Differences in GI values resulted in marginally significant differences between RU-DIMESA and COMM (difference 21.4; 95% C.I. −2~4.7; p-value = 0.092) and between RU-PONTE and ICS (difference 22.1; 95% C.I. −0.8~45.0; p-value = 0.092).

For each pasta sample, the ICS, COMM, RU-DIMESA RU-PONTE, TI-DIMESA, and TI-PONTE, and for each reference food session, data of the average glycemic response (mmol /L) after food ingestion, recorded at zero (t0), 15, 30, 45, 60, 90, and 120 min, were reported (Tables 6 and 7).

Table 5. Pairwise comparison of glycemic index among all pasta samples.

Two Comparison Samples	Difference	Lower 95% CI	Upper 95% CI	*p*-Value
ICS-COMM	−3.7	−27.0	19.6	0.997
RU DIMESA-COMM **	**21.4**	**−2.0**	**44.7**	**0.092**
RU PONTE-COMM	18.4	−4.9	41.7	0.205
TI DIMESA-COMM	19.3	−4.0	42.6	0.164
TI PONTE-COMM *	**25.3**	**2.0**	**48.6**	**0.026**
RU DIMESA-ICS *	**25.1**	**2.2**	**48.0**	**0.024**
RU PONTE-ICS **	**22.1**	**−0.8**	**45.0**	**0.065**
TI DIMESA-ICS *	**23.0**	**0.1**	**45.9**	**0.049**
TI PONTE-ICS *	**29.0**	**6.1**	**51.9**	**0.005**
RU PONTE-RU DIMESA	−3.0	−25.9	19.9	0.999
TI DIMESA-RU DIMESA	−2.1	−25.0	20.8	1.000
TI PONTE-RU DIMESA	3.9	−19.0	26.8	0.996
TI DIMESA-RU PONTE	0.9	−22.0	23.8	1.000
TI PONTE-RU PONTE	6.9	−16.0	29.8	0.950
TI PONTE-TI DIMESA	6.0	−16.9	28.9	0.972

C.I.= Confidence Interval; * Marginal significant difference; ** Significant difference.

Table 6. For pasta sample ICS, COMM, RU-DIMESA RU-PONTE, TI-DIMESA, TI-PONTE tables (a–f) report the average glycemic response (mmol /L) after food ingestion, recorded at zero (t0), 15, 30, 45, 60, 90, and 120 min in enrolled volunteers; SD = Standard deviation; Var = Variance.

ICS (a)				COMM (b)			
T *Minutes*	Glyc *Mean*	SD	Var	T *Minutes*	Glyc *Mean*	SD	Var
0	5.26	0.33	0.11	0	5.20	0.55	0.31
15	5.79	0.58	0.34	15	5.80	0.59	0.35
30	6.35	0.51	0.26	30	6.31	0.78	0.62
45	5.97	0.57	0.33	45	5.86	0.69	0.47
60	5.8	0.71	0.51	60	5.90	0.93	0.87
90	5.81	0.49	0.24	90	5.58	0.49	0.24
120	5.65	0.51	0.26	120	5.42	0.41	0.17
RU-DIMESA (c)				**RU-PONTE (d)**			
T *Minutes*	Glyc *Mean*	SD	Var	T *Minutes*	Glyc *Mean*	SD	Var
0	4.90	0.36	0.13	0	5.30	0.35	0.12
15	5.57	0.58	0.34	15	5.90	0.53	0.28
30	6.69	0.79	0.63	30	7.34	0.49	0.24
45	6.64	0.83	0.69	45	6.92	1.09	1.19
60	6.18	0.97	0.94	60	6.32	0.94	0.88
90	5.72	0.65	0.43	90	5.81	0.56	0.31
120	5.56	0.57	0.32	120	5.61	0.60	0.36
TI-DIMESA (e)				**TI-PONTE (f)**			
T *Minutes*	Glyc *Mean*	SD	Var	T *Minutes*	Glyc *Mean*	SD	Var
0	5.15	0.5	0.25	0	4.82	0.43	0.18
15	5.78	0.67	0.45	15	5.68	0.45	0.21
30	6.74	0.87	0.76	30	6.61	0.64	0.41
45	6.46	0.89	0.79	45	6.33	0.96	0.92
60	6.25	0.72	0.52	60	5.9	0.89	0.79
90	5.93	0.56	0.31	90	5.74	0.81	0.66
120	5.94	0.46	0.21	120	5.53	0.66	0.44

Table 7. For reference food (50 g of glucose in 250 mL mineral water) tables (g,h) report the average values recorded for glycemia (mmol/L), SD (Standard deviation), and Var (Variance) in enrolled volunteers. Reference food was tested twice in two different sessions (Reference 1/Reference 2).

REFERENCE 1 (g)				REFERENCE 2 (h)			
T *Minutes*	Glyc *Mean*	SD	Var	T *Minutes*	Glyc *Mean*	SD	Var
0	5.16	0.43	0.18	0	5.19	0.38	0.15
15	6.82	0.98	0.97	15	6.79	1.03	1.07
30	8.38	1.31	1.71	30	8.21	1.25	1.57
45	8.66	1.53	2.35	45	8.02	1.51	2.29
60	8.24	1.45	2.11	60	7.67	1.08	1.16
90	6.03	1.13	1.29	90	5.45	1.05	1.11
120	5.06	1.24	1.53	120	4.65	0.93	0.87

The analysis of variance of the glycemic indexes showed a significant difference in the mean values for the different types of pasta (p-value < 0.001; Table 8).

Table 8. One-way ANOVA test (six pasta sessions and two reference food sessions).

Time	p-Value
0	0.095
15	**<0.001**
30	**<0.001**
45	**<0.001**
60	**<0.001**
90	0.743
120	**0.002**

The analysis of variance (ANOVA) showed that, excluding time at 0 (T0) and 90 min (T90), there was always a significant difference between the mean values of glycemia in each sampling time (six pasta samples and two reference foods).

Post-hoc pairwise comparison analysis showed statistically significant higher glycemia values of the RU-PONTE versus COMM (difference +1.04 mmol; 95% C.I. 0.25~1.82; p-value = 0.003) and versus ICS (difference +0.99 mmol; 95% C.I. 0.23~1.76; p-value = 0.003) at 30 min (T30); difference between RU-PONTE and TI-PONTE was marginally significant (difference +0.74 mmol; 95% C.I. −0.03~1.50; p-value = 0.067). RU-PONTE glycemic values resulted in being statistically higher with respect to COMM (difference +1.06 mmol; 95% C.I. 0.09~2.03; p-value = 0.023) and ICS (difference +0.96 mmol; 95% C.I. 0.01~1.91; p-value = 0.047) at 45 min (T45).

The two reference food curves did not show any significant differences at every time point.

4. Discussion

The results obtained from the study revealed a lower GI for industrial pasta ICS and COMM (34.1 and 37.8) and higher GIs for handcrafted landraces (*Timilia* and *Russello*) pasta (RU-PONTE, RU-DIMESA, TI-PONTE e TI_DIMESA). By applying the standard GI measurement methodology, this study produced unique data for pasta made from durum wheat landraces cultivated in Sicily. The results do not support the popular idea of a reduced glycemic response elicited by *Timilia* and *Russello* wheat landrace pasta. Further investigations are needed to clarify the supposed healthier value of "ancient grain" based food.

The meager data available for GI of Italian pasta [14] revealed that values higher than 55 were not common for durum wheat pasta made without other ingredients and additives. On the other hand, low values (below 35) were mainly expected for wholemeal pasta, with some exceptions. Our results were partially in line with these data.

The four samples obtained from Sicilian landraces (RU-PONTE, RU-DIMESA, TI-PONTE e TI_DIMESA) showed values ranging from 56.2 to 63.1. Although these samples were made with wholemeal flour, usually associated with lower GI [25–28], a higher GI was recorded. It is known that the presence of fiber may result in an unstructured pasta matrix, leading to a more rapid digestion of starch and, subsequently, to higher GI [29,30]. These data confirm that it is necessary to clarify the role of the fiber content in reducing GI, in moderating the rate of glucose absorption and controlling the glucose response curve.

It is hypothesized that other process variables make the GI higher with a more relevant impact with respect to the fiber content. Among these variables could be considered the fine grain size, which characterizes the stone-milled flour [5–8], the low-temperature used for drying processes [31], the low gluten index that characterizes Sicilian landraces [19], and the probable presence of soft grains in the samples of Sicilian landraces [32]. All these variables contribute to making GI higher.

The stone-milled flours used to produce experimental or commercial samples of pasta from *Timilia* and *Russello* landraces were characterized by a large fraction of *"farinette"* (more than 20% of particles of flour with a diameter $\leq$118 μ), and it could be related to a higher GI and may have a relevant impact on starch availability to be depolymerized [5–8]. Indeed, the traditional stone milling of grains can consistently damage the cell wall structure, increasing the exposure of the entrapped starch granules to enzymatic digestion [7,8,33].

The stone milling process is prone to damage the cell walls, which may have in turn affected the gelatinization degree of starch granules during cooking [26,34–36]. The degree of starch damage resulting from milling is one of the factors known to increase the starch gelatinization degree during wet cooking [37–41].

The results about pasta ICS can be discussed by considering the variables that produced this type of experimental sample, which showed the lowest GI (34.1) compared to the other samples of the study, which were all wholemeal.

Pasta ICS was an experimental pasta made of a mix of modern semolina wheat varieties whose peculiar characteristic was the total removal of the *"farinette"* fraction. The steel roll milling of the wheat varieties used for pasta ICS, instead, gave mixed semolina characterized by larger and (probably) less damaged particles, a trait magnified by the total experimental removal of the *farinette* fraction.

The comparable sample named COMM, a wholemeal commercial Italian pasta (spaghetti) showed a slightly higher GI value despite its fiber content (37.8).

For the pasta COMM sample, a low GI value would be expected due to its high fiber content.

Fiber content, an otherwise healthy and recommended element of the diet, in this study seems to not be strongly related to the GI values of the samples. It should also be considered that the total dietary fiber content for the samples was estimated without discriminating between the insoluble/soluble fractions, which might have had a different impact on the GI of food [25–28].

There could be a relationship, instead, between the grinding diagrams of flours or semolina used to make pasta and the GI results of the samples. In further studies, the role of stone milling processes versus steel roll milling could be compared in order to evaluate the impact on GI of these different grinding methodologies.

There are at least two other factors we may consider relevant to determining a low GI for pasta ICS, compared to the intermediate GI for *Timilia* and *Russello* pasta samples, namely the drying temperature diagram and the wheat varieties used.

For traditional Sicilian landraces, *Timilia,* and *Russello* pasta samples, the protein content recorded was around 10% (see Table 2), whilst it was over 12% for pasta ICS and pasta COMM. Sicilian landraces are known for being characterized by a lower gluten index (a widely accepted indicator of gluten strength) if compared to "modern" varieties [19]. The different protein content and mostly gluten strength are important factors for determining the tighter or less tight interaction between the starchy matrix and the gluten net in the dough and in the final product [29,39].

This aspect may have been amplified by the low-temperature (less than 40 °C) drying process submitted by *Timilia* and *Russello* pasta samples, resulting in a different texture of pasta and in a leakier interaction of starch with the gluten net.

On the other hand, drying pasta at high temperature (90 °C), as it was conducted for pasta ICS and probably pasta COMM, contributed to protein aggregation and to further strengthen the protein network around the starch, reducing its digestibility [29].

The gluten weakness for the landrace's pasta samples could be explained by the releasing of a greater amount of starch in the cooking water compared to the ICS and COMM samples (see Table 3), supporting the hypothesis for Sicilian landraces wheat pastas of a starch matrix less effectively trapped in the gluten net. Indeed, the number of servings for different samples needed to be adjusted after the preliminary cooking tests.

Another reason to explain the higher GI values recorded for Sicilian landrace pasta samples may be found in the accidental and common contamination of *Timilia* and *Russello* crops with a certain amount of soft wheat due to its presence in seed grains and in saw bed [23]. The higher ratio amylopectin/amylose, typical of soft grain-derived starches, could partially explain the intermediate GI values recorded for landrace pasta [12,13].

In summary, according to the collected data and the study results, it was hypothesized that each one of the different pasta production steps, i.e., the milling procedures of the grain, the drying temperature diagram and the choice of a specific wheat variety, should be considered relevant in affecting the GI values of the final products.

5. Conclusions

The commonly held belief that ancient grains hold a higher health value than conventional or modern variety grain chains is not supported in this GI study involving pasta samples, regardless whether these products are handcrafted, stone-milled wholemeal flours. Pasta samples obtained from landraces always showed a higher GI value if compared to other experimental or commercial samples.

However, it is plausible to hypothesize that by acting in a controlled manner on specific process variables, the result of the production of a low GI type of pasta can be pursued. It is possible to act on production processes, to control variables as wheat varietal selection, cultivation techniques, gluten strength with the protein content, grinding diagram and typology of the milling process, extrusion pressure [42] of the pasta production and shaping, in order to effectively reduce the GI of pasta, and maybe other wheat derivatives, without altering their original formula by the addition of extra ingredients.

In this contest, however, it is necessary to clarify which process variables have major impacts on determining a lower glycemic response and a low GI in order to transfer to the pasta factories the technological protocols for the production of innovative low-GI pasta. The fact that the presence of dietary fiber does not always correspond to a lowering of the GI value has also been confirmed.

Greater efforts must be made to understand how the choice of a grinding diagram that eliminates "farinette" fraction from semolina, or the use of high temperatures in the drying processes, can lead to a lowering of the GI.

Author Contributions: Conceptualization, G.R. and A.P.; methodology, A.P.; validation, G.R. and P.A; formal analysis, A.P. and G.R.; investigation, A.P. and G.R.; resources, G.R.; data curation, A.P.; writing—original draft preparation, A.P.; writing—review and editing, G.R.; visualization, B.M.; supervision, G.R.; project administration, G.R.; funding acquisition, G.R. All authors have read and agreed to the published version of the manuscript.

Funding: This research was funded with Project DiMeSa that has been financed by a grant (PON02 00667-PON02 00451 3361785) of the Italian Ministry of University and Research (MIUR).

Institutional Review Board Statement: The study was conducted according to the guidelines of the Declaration of Helsinki and approved by the Ethics Committee PALERMO 2 of AOR VILLA SOFIA CERVELLO -ARNAS CIVISO–ISMETT-ASP TP-ASP AG (protocol code CIVICO/19 approval on the 15 January 2015).

Informed Consent Statement: Informed consent was obtained from all subjects involved in the study.

Data Availability Statement: The datasets generated for this study are available on request to the corresponding author.

Acknowledgments: The authors are grateful to Norberto Pogna for his priceless professional work as President of the Ballatore Research Consortium during the time of the project DIMESA. We also wish to thank Tomasello Pasta factory in Casteldaccia (Palermo, Italy) for its precious collaborative efforts in preparation of samples. We are grateful to nurses Tinervia Marco, Stefania Leggio and Marco Vincenzo Calderone, for their professional management of glycemic measurements. Lastly, we are grateful to Marco Ventimiglia for his precious and professional support and suggestions for statistical analysis.

Conflicts of Interest: The authors declare no conflict of interest. The funders had no role in the design of the study; in the collection, analyses, or interpretation of data, in the writing of the manuscript, or in the decision to publish the results.

References

1. Jenkins, D.J.; Wolever, T.M.; Taylor, R.H.; Barker, H.; Fielden, H.; Baldwin, J.M.; Bowling, A.C.; Newman, H.C.; Jenkins, A.L.; Goff, D.V. Glycemic index of foods: A physiological basis for carbohydrate exchange. *Am. J. Clin. Nutr.* **1981**, *34*, 362–366. [CrossRef] [PubMed]
2. Wolever, T.M.S. *The Glycaemic Index—A Physiological Classification of Dietary Carbohydrate*; CABI: Wallingford, UK, 2006; ISBN 1-84593-051-7.
3. Brouns, F.; Bjorck, I.; Frayn, K.N.; Gibbs, A.L.; Lang, V.; Slama, G.; Wolever, T.M.S. Glycaemic index methodology. *Nutr. Res. Rev.* **2005**, *18*, 145–171. [CrossRef] [PubMed]
4. FAO (Food and Agriculture Organization); WHO (World Health Organization). *Carbohydrates in Human Nutrition*; Report of a Joint FAO/WHO Expert Consultation 1998; FAO Food and Nutrition Paper 66; FAO: Rome, Italy, 1998.
5. Ramdath, D.D. Glycemic Index, Glycemic Load, and Their Health Benefits. In *Reference Module in Food Science*; Elsevier: Amsterdam, The Netherlands, 2016.
6. Sadler, M. *Food, Glycaemic Response and Health*; ILSI Europe a.i.s.b.l.: Brussels, Belgium, 2011; 34p.
7. Dhital, S.; Bhattarai, R.R.; Gorham, J.; Gidley, M.J. Intactness of cell wall structure controls the in vitro digestion of starch in legumes. *Food Funct.* **2016**, *7*, 1367–1379. [CrossRef]
8. Al-Rabadi, G.J.; Gilbert, R.G.; Gidley, M.J. Effect of particle size on kinetics of starch digestion in milled barley and sorghum grains by porcine alpha-amylase. *J. Cereal Sci.* **2009**, *50*, 198–204. [CrossRef]
9. Augustin, L.S.A.; Kendall, C.W.C.; Jenkins, D.J.A.; Willett, W.C.; Astrup, A.; Barclay, A.W.; Björck, I.; Brand-Miller, J.C.; Brighenti, F.; Buyken, A.E.; et al. Glycemic index, glycemic load and glycemic response: An International Scientific Consensus Summit from the International Carbohydrate Quality Consortium (ICQC). *Nutr. Metab. Cardiovasc. Dis.* **2015**, *25*, 795–815. [CrossRef] [PubMed]
10. Jenkins, D.J.; Kendall, C.W.; Augustin, L.S.; Franceschi, S.; Hamidi, M.; Marchie, A.; Jenkins, A.L.; Axelsen, M. Glycemic index: Overview of implications in health and disease. *Am. J. Clin. Nutr.* **2002**, *76*, 266S–273S. [CrossRef] [PubMed]
11. Venn, B.J.; Green, T.J. Glycemic index and glycemic load: Measurement issues and their effect on diet–disease relationships. *Eur. J. Clin. Nutr.* **2007**, *61* (Suppl. 1), S122–S131. [CrossRef]
12. Atkinson, F.S.; Foster-Powell, K.; Brand-Miller, J.C. International Tables of Glycemic Index and Glycemic Load Values: 2008. *Diabetes Care* **2008**, *31*, 2281–2283. [CrossRef]
13. Foster-Powell, K.; Holt, S.H.; Brand-Miller, J.C. International table of glycemic index and glycemic load values: 2002. *Am. J. Clin. Nutr.* **2002**, *76*, 5–56. [CrossRef]
14. Scazzina, F.; Dall'Asta, M.; Casiraghi, M.C.; Sieri, S.; Del Rio, D.; Pellegrini, N.; Brighenti, F. Glycemic index and glycemic load of commercial Italian foods. *Nutr. Metab. Cardiovasc. Dis.* **2016**, *26*, 419–429. [CrossRef]
15. Van den Broeck, H.C.; De Jong, H.C.; Salentijn, E.M.J.; Dekking, L.; Bosch, D.; Hamer, R.J.; Gilissen, L.J.W.J.; Van der Meer, I.M.; Smulders, M.J.M. Presence of celiac disease epitopes in modern and old hexaploid wheat varieties: Wheat breeding may have contributed to increased prevalence of celiac disease. *Theor. Appl. Genet.* **2010**, *121*, 1527–1539. [CrossRef]
16. Malagolda, M.; Meinhardt, S.W.; Simsek, S. Detection and quantitation of immunogenetic epitopes related to celiac disease in historical and modern hard red spring wheat cultivar. *Food Chem.* **2018**, *264*, 101–107. [CrossRef]
17. Ficco, D.B.M.; Prandi, B.; Amaretti, A.; Anfelli, I.; Leonardi, A.; Raimondi, S.; Pecchioni, N.; De Vita, P.; Faccini, A.; Sforza, S.; et al. Comparison of gluten peptides and potential prebiotic carbohydrates in old and modern *Triticum turgidum* ssp. Genotypes. *Food Res. Int.* **2019**, *120*, 568–576. [CrossRef]

18. Ribeiro, M.; Rodriguez-Quijano, M.; Nunes, F.M.; Carrillo, J.M.; Branlard, G.; Igrejas, G. New insights into wheat toxicity: Breeding did not seem to contribute to a prevalence of potential celiac disease's immunostimulatory epitopes. *Food Chem.* **2016**, *213*, 8–18. [CrossRef] [PubMed]
19. De Santis, M.A.; Giuliani, M.M.; Giuzio, L.; De Vita, P.; Lovegrove, A.; Shewry, P.R.; Flagella, Z. Differences in gluten protein composition between old and modern durum wheat genotypes in relation to 20th century breeding in Italy. *Eur. J. Agron.* **2017**, *87*, 19–29. [CrossRef]
20. Dinu, M.; Whittaker, A.; Pagliai, G.; Benedettelli, S.; Sofi, F. Ancient wheat species and human health: Biochemical and clinical implications. *J. Nutr. Biochem.* **2018**, *52*, 1–9. [CrossRef] [PubMed]
21. Shewry, P.R. Do ancient types of wheat have health benefits compared with modern bread wheat? *J. Cereal Sci.* **2018**, *79*, 469–476. [CrossRef]
22. Freckmann, G.; Schmid, C.; Baumstark, A.; Pleus, S.; Link, M.; Haug, C. System Accuracy Evaluation of 43 Blood Glucose Monitoring Systems for Self-Monitoring of Blood Glucose according to DIN EN ISO 15197. *J. Diabetes Sci. Technol.* **2012**, *6*, 1060–1075. [CrossRef]
23. D'Egidio, M.G.; Nardi, S. Textural measurement of cooked spaghetti. In *Pasta and Noodle Technology*; Kruger, J.E., Matsuo, R.R., Dick, J.W., Eds.; AACC, Inc.: St. Paul, MN, USA, 1996; pp. 133–156.
24. Ente Nazionale Italiano di Unificazione. *Semola e Semolato di Grano Duro—Determinazione della Granulometria*; UNI 10.873: 2000; Ente Nazionale Italiano di Unificazione: Roma, Italy, 2000.
25. Schwartz, S.E.; Levine, G.D. Effects of dietary fiber on intestinal glucose absorption and glucose tolerance in rats. *Gastroenterology* **1980**, *79 Pt 1*, 833–836. [CrossRef]
26. Goff, H.D.; Repin, N.; Fabek, H.; El Khoury, D.; Gidley, M.J. Dietary fibre for glycaemia control: Towards a mechanistic understanding. *Bioact. Carbohydr. Diet. Fibre* **2018**, *14*, 39–53. [CrossRef]
27. Raben, A.; Tagliabue, A.; Christensen, N.J.; Madsen, J.; Holst, J.J.; Astrup, A. Resistant starch: The effect on postprandial glycemia, hormonal response, and satiety. *Am. J. Clin. Nutr.* **1994**, *60*, 544–551. [CrossRef]
28. Kimura, T. The regulatory effects of resistant starch on glycaemic response in obese dogs. *Arch. Anim. Nutr.* **2013**, *67*, 503–509. [CrossRef] [PubMed]
29. Gallo, V.; Romano, A.; Masi, P. Does the presence of fibres affect the microstructure and in vitro starch digestibility of commercial Italian pasta. *Food Struct.* **2020**, *24*, 100139–100145. [CrossRef]
30. Dodi, R.; Bresciani, L.; Biasini, B.; Cossu, M.; Scazzina, F.; Taddei, F.; D'Egidio, M.G.; Dall'Asta, M.; Martini, D. Traditional and Non-Conventional Pasta-Making Processes: Effect on In Vitro Starch Digestibility. *Foods* **2021**, *10*, 921. [CrossRef]
31. Kim, E.H.-J.; Petrie, J.R.; Motoi, L.; Morgenstern, M.P.; Sutton, K.H.; Mishra, S.; Simmons, L.D. Effect of structural and physicochemical characteristics of the protein matrix in pasta on in vitro starch digestibility. *Food Biophys.* **2008**, *3*, 229–234. [CrossRef]
32. Russo, G. Frumenti Siciliani: Qualità e valore nutrizionale. In *I Grani Duri Siciliani. Storia, Antropologia, Gastronomia*; Nocilla, N., Silvestri, G., Eds.; Kalos: Palermo, Italy, 2018; pp. 131–150.
33. SMiles, M.J.; Morris, V.J.; Orford, P.D.; Ring, S.G. The roles of amylose and amylopectin in the gelation and retrogradation of starch. *Carbohydr. Res.* **1985**, *135*, 271–281.
34. Jung, E.Y.; Suh, H.J.; Hong, W.S.; Kim, D.G.; Hong, Y.H.; Hong, I.S.; Chang, U.J. Uncooked rice of relatively low gelatinization degree resulted in lower metabolic glucose and insulin responses compared with cooked rice in female college students. *Nutr. Res.* **2009**, *29*, 457–461. [CrossRef] [PubMed]
35. Parada, J.; Aguilera, J.M. In vitro digestibility and glycemic response of potato starch is related to granule size and degree of gelatinization. *J. Food Sci.* **2009**, *74*, E34–E38. [CrossRef]
36. Wang, S.; Wang, S.; Liu, L.; Wang, S.; Copeland, L. Structural orders of wheat starch do not determine the in vitro enzymatic digestibility. *J. Agric. Food Chem.* **2017**, *65*, 1697–1706. [CrossRef] [PubMed]
37. Kong, B.W.; Kim, J.I.; Kim, M.J.; Kim, J.C. Porcine pancreatic α-amylase hydrolysis of native starch granules as a function of granule surface area. *Biotechnol. Prog.* **2003**, *19*, 1162–1166. [CrossRef]
38. Planchot, V.; Colonna, P.; Buleon, A.; Gallant, D. Amylolysis of starch granules and α-glucan crystallites. In *Starch Structure and Functionality*; Frazier, P.J., Donald, A.M., Richmond, P., Eds.; The Royal Society of Chemistry: London, UK, 1997; pp. 141–152.
39. Ring, S.G.; Gee, J.M.; Whittam, M.; Orford, P.; Johnson, I.T. Resistant starch: Its chemical form in foodstuffs and effect on digestibility in vitro. *Food Chem.* **1988**, *28*, 97–109. [CrossRef]
40. Hu, P.; Zhao, H.; Duan, Z.; Linlin, Z.; Wu, D. Starch digestibility and the estimated glycemic score of different types of rice differing in amylose contents. *J. Cereal Sci.* **2004**, *40*, 231–237. [CrossRef]
41. Salman, H.; Copeland, L. Effect of repeated heating and cooling cycles on the pasting properties of starch. *J. Cereal Sci.* **2009**, *51*, 105–109. [CrossRef]
42. Petitot, M.; Brossard, C.; Barron, C.; Larré, C.; Morel, M.-H.; Micard, V. Modification of pasta structure induced by high drying temperatures. Effects on the in vitro digestibility of protein and starch fractions and the potential allergenicity of protein hydrolysates. *Food Chem.* **2009**, *116*, 401–412. [CrossRef]

foods

MDPI

Article

In Vivo and In Vitro Starch Digestibility of Fresh Pasta Produced Using Semolina-Based or Wholemeal Semolina-Based Liquid Sourdough

Simonetta Fois, Piero Pasqualino Piu, Manuela Sanna, Tonina Roggio and Pasquale Catzeddu *

Porto Conte Ricerche Srl, Località Tramariglio, 07041 Alghero, Italy; fois@portocontericerche.it (S.F.); piupietro@gmail.com (P.P.P.); sanna@portocontericerche.it (M.S.); roggio@portocontericerche.it (T.R.)
* Correspondence: catzeddu@portocontericerche.it; Tel.: +39-079-998400

Abstract: The use of wholemeal flour and sourdough fermentation in different food matrices has received considerable attention in recent years due to its resulting health benefits. In this study, a semolina-based and a wholemeal semolina-based sourdough were prepared and added to the formulation of gnocchetti-type fresh pasta. Four types of gnocchetti were made, using semolina plus semolina-based sourdough (SS), semolina plus wholemeal semolina-based sourdough (SWS), semolina alone (S), and semolina plus wholemeal semolina (WS). The latter two were used as controls. The digestibility of starch was studied both in vitro and in vivo, and the glycemic response (GR) and glycemic load (GL) were determined. Starch digestibility, both in vivo and in vitro, was higher in wholemeal semolina than semolina pasta and the resulting GR values (mg dL^{-1} min^{-1}) were also higher (2209 and 2277 for WS and SWS; 1584 and 1553 for S and SS, respectively). The use of sourdough significantly reduced the rapidly digestible starch (RDS) content and increased the inaccessible digestible starch (IDS) content. The addition of sourdough to the formulation had no effect on the GR values, but led to a reduction of the GL of the pasta. These are the first data on the GR and GL of fresh pasta made with sourdough.

Keywords: starch; available carbohydrates; glycemic response; glycemic index; glycemic load

Citation: Fois, S.; Piu, P.P.; Sanna, M.; Roggio, T.; Catzeddu, P. In Vivo and In Vitro Starch Digestibility of Fresh Pasta Produced Using Semolina-Based or Wholemeal Semolina-Based Liquid Sourdough. *Foods* **2021**, *10*, 2507. https://doi.org/10.3390/foods10102507

Academic Editor: Arun K. Bhunia

Received: 29 June 2021
Accepted: 15 October 2021
Published: 19 October 2021

1. Introduction

Pasta is a staple food, highly appreciated and consumed in several countries. In 2019 the world production of pasta reached almost 16 million tons and Italy was the leading pasta producer (3.5 million tons). Six Italians out of ten eat pasta on a daily basis, amounting to about 23.1 kg per capita per year of pasta consumed [1]. Dried pasta (less than 12.5% moisture content) represents the main global market share, while fresh pasta (more than 24.0% moisture content) represents a small but increasing share, particularly in the Italian market.

Pasta is commonly produced from durum wheat semolina but the addition of flours from other raw materials, such as rice, buckwheat, barley, spelt, millet, oats, quinoa, and legumes, has been investigated for the purpose of improving the nutritional quality and the health-promoting effects of pasta [2]. Many studies have investigated the effect of pasta fortified by a wide range of supplements, some of which, such as dietary fibers, amino acids, peptides and vitamins, can be considered functional ingredients that give pasta a potentially positive effect on health [3]. The consumption of fiber-rich pasta, or wholemeal pasta, has increased over the past few years, probably since it has well-recognized beneficial effects on human health, such as preventing cardiovascular disease, obesity, cancer, and diabetes risk factors [4,5]. Moreover, consumption of cereal fibers can increase the sensation of satiety and fullness [6], gut microbial diversity and abundance [7], and can also reduce the glycemic index, as observed in wholemeal bread by Scazzina et al. [8]. Unfortunately, decay of sensory and textural characteristics is observed when bran or wholemeal flour are incorporated into the foods [9], which reduces the demand for fiber-rich foods. To overcome these barriers, use of sourdough technology for wholemeal bread has been investigated in depth and has been reported to improve the sensory, nutritional and health quality of the bread itself. Fermented bran was successfully used to improve the sensory properties of bread containing bran [10]. An improvement in the sensory properties of fiber-rich pasta was observed when fermented wholemeal semolina was used [11].

Pasta is a good source of carbohydrates and the major dietary source of energy in the Mediterranean diet. It is considered a low glycemic index (GI) food [12], as glucose is liberated from carbohydrates fairly slowly after ingestion. The low GI of pasta is mainly due to its dense structure. This results in slow digestion and delayed gastric emptying, making pasta a unique example of a refined cereal food with a low GI. Pasta consumed in the context of a low-GI diet allows body weight to be reduced [13] and has a positive effect against diabetes risk factors [14] since it prevents the high postprandial glucose and insulin peaks, which may contribute to the development of insulin resistance and type 2 diabetes. In people with type-2 diabetes, consumption of pasta without exceeding the limits recommended for total carbohydrate intake, is not associated with a worsening of obesity or cardiovascular risk factors [12].

Sourdough fermentation has been shown to decrease the glycemic index in foods [8]. Use of sourdough technology has been investigated in fresh pasta made with fermented semolina and fermented wholemeal semolina [11,15], and data on the physical, chemical, and sensory properties of the pasta have been reported.

The aim of this work was to evaluate the effect of fermentation technology and the addition of wholemeal semolina, and the effect of their interaction on the in vitro digestion of carbohydrates and in vivo glycemic response of fresh pasta.

2. Materials and Methods

2.1. Raw Materials

Commercial wholemeal semolina (Integrale, Selezione Casillo S.r.l., Corato, Bari, Italy), and commercial semolina (Extra Arancio, Selezione Casillo S.r.l., Corato, Bari, Italy) were used. The percentage composition of the wholemeal semolina, as is or on a dry matter basis (D.M.), was: moisture 14.1%, ash 1.6% D.M., protein 12.5% D.M., fiber 7.8% D.M., dry gluten 8.5% D.M., gluten index 60%, alveographic W 199 (J $\times$ 10^{-4}) and P to L ratio 5.12. The composition of semolina was: moisture 14.0%, ash 0.75% D.M., protein 13% D.M., fiber 2.7% D.M., dry gluten 11% D.M., gluten index 88%, alveographic W 176 (J $\times$ 10^{-4}) and P to L ratio 1.31.

2.2. Preparation and Maintenance of Liquid Sourdough

The liquid sourdough starter was prepared according to Fois et al. [11] and was refreshed using a 5 L sourdough maker machine (Starpizza S.A.S., Verona, Italy). Two different sourdoughs were prepared: the first was a semolina-based sourdough prepared using water and semolina, while the second was a wholemeal semolina-based sourdough prepared with water and wholemeal semolina. Both were refreshed by daily back-slopping during which sourdough, water and semolina or wholemeal semolina were mixed in a ratio of 1:1:1, to obtain a dough yield of 200. The semolina-based sourdough had a pH value of 4.4 and a TTA of 10.0 mL NaOH (0.1 mol/L) in 10 g; the wholemeal-based sourdough had a pH value of 4.2 and a TTA of 13.0 mL NaOH (0.1 mol/L) in 10 g.

2.3. Fresh Pasta Making

Fresh pasta (of the gnocchetti sardi type) was prepared using the La Monferrina Dolly pasta maker (La Monferrina, Moncalieri, Italy) equipped with a bronze die. Four different pasta formulations were prepared:

Sample S: pasta made with semolina (1000 g) and water (300 g).

Sample SS: pasta made with semolina (700 g) and semolina-based sourdough (600 g). No extra water was added since the sourdough contained 50% of water.

Sample WS: pasta made with wholemeal semolina (300 g), semolina (700 g) and water (300 g).

Sample SWS: pasta made with whole meal semolina-based sourdough (600 g) and semolina (700 g). No extra water was added, as in sample SS.

After production, the fresh pasta was immediately pasteurized and packaged under modified atmosphere (CO_2:N_2 = 30:70), as in Fois et al. [15]. It was then stored at 4 °C until analysis.

2.4. Chemical Analysis of Cooked Pasta

The moisture content of the pasta was measured at 105 °C with a Thermogravimetric Analyzer Thermostep (Eltra GmbH, Haan, Germany). Total titratable acidity (TTA) and pH were determined with an automatic titrator (Crison, Hach Lange, Barcelona, Spain), after homogenizing a 10 g sample in 90 mL of distilled water. After 30 min of gentle stirring for sourdough and 60 min for chopped and homogenized pasta, the pH was determined and the samples were titrated to pH 8.5 with NaOH 0.1 mol/L. The TTA was expressed as mL of NaOH per 10 g of sample. Available carbohydrates (ACH) were determined (g/100 g of cooked pasta) using the Available Carbohydrates Assay Kit (Megazyme, Wicklow, Ireland). All the data have been reported in Table 1.

Table 1. Chemical properties of cooked pasta.

Pasta	Moisture (g/100 g)	pH	TTA [1] (mL NaOH N/10)	Available Carbohydrates [2]
S	57.48 ± 0.59	6.53 ± 0.07	0.85 ± 0.03	39.24 ± 1.34
SS	56.76 ± 0.44	5.52 ± 0.17	2.01 ± 0.44	33.28 ± 1.69
WS	58.32 ± 2.24	6.61 ± 0.02	1.24 ± 0.06	34.27 ± 0.82
SWS	55.85 ± 1.13	5.79 ± 0.04	2.32 ± 0.01	31.32 ± 0.58

(S, pasta with semolina; SS, pasta with semolina-based sourdough; WS, pasta with wholemeal semolina; SWS, pasta with wholemeal semolina-based sourdough. Mean value of at least four replicates ± standard deviation. [1] TTA, total titratable acidity expressed as mL of 0.1 N NaOH/10 g of pasta dry matter. [2] Grams of glucose in 100 g of cooked pasta, as is basis).

2.5. In Vitro Starch Digestibility

The pasta was cooked for the optimum cooking time and was then roughly chopped with a knife in order to simulate chewing. In vitro digestion was then performed as for Sanna et al. [16]. Digested samples were collected at 20, 60, 90, 120, and 180 min and the hydrolysis curves were built. Digestion was carried out in triplicate. The area under the hydrolysis curves (0–180 min) was calculated and the hydrolysis index (HI) was obtained as the ratio between the area under the curve of the sample and the area under the curve of the reference food (white breadcrumbs). The estimated glycemic index (GIe) was calculated following the Equation (1), as in Fico et al. [17]:

$$GIe = 8.198 + 0.862 \times HI \quad (1)$$

Rapidly digestible starch (RDS), slowly digestible starch (SDS) and inaccessible digestible starch (IDS) were calculated. RDS is the glucose released after 20 min of in vitro digestion. SDS and IDS are defined as the glucose released within the 20 and 120 min time-frame and within the 120 and 180 min time-frame, respectively. IDS is defined as "inaccessible digestible starch" since it is not actually digestion-resistant starch but just physically inaccessible to the digestive enzymes. It was made accessible by homogenizing the sample after 120 min of in vitro digestion [16].

2.6. Glycemic Response (GR) and Glycemic Load (GL)

Glycemic response was determined in vivo following the procedures reported by the Food and Agriculture Organization/World Health Organization [18] and the method reported by Sugiyama et al. [19]. The study was approved by the Sardinian Ethical Committee of the Azienda Tutela Salute (ATS). Fifteen volunteers (seven female and eight male) were recruited by the Porto Conte Ricerche laboratory (Alghero, Italy), where the study was conducted. Volunteers aged between 25 and 55 years old were chosen on the basis of their body mass index (BMI), which had to range between 18.5 and 24.9 kg m^{-2}. BMI was calculated as the ratio of weight (kg) to squared height (m), as indicated in: http://www.salute.gov.it/portale/salute/p1_5.jsp?id=135&area=Vivi_sano (accessed on 21 December 2020).

The volunteers were informed about the main criteria required in order to be eligible to take part in the study (i.e., they had to be healthy, not affected by chronic diseases, free from food intolerances and allergies, not involved in drug usage, not in a special physiological condition, such as breastfeeding) and what they would be asked to do. The volunteers were then asked to formalize their membership by signing an agreement. Obviously, each one was free to withdraw from the experimentation at any time without explanation.

2.6.1. Experimental Procedures

The pasta was cooked for its optimum cooking time, according to the AACC Approved Method 66-50 [20], in previously salted (1.3% *w*/*v*) boiling water. The ratio of pasta to cooking water was 25 g per 300 mL. The pasta was served as is, with no seasoning. The four types of pasta were served to the volunteers on different days, in a randomized order so that none of the volunteers exceeded one meal per two weeks, and were tested for equivalent available carbohydrate content (50 g). Monohydrate glucose solution (Glucose (Monico) 50% p/V, Monico SPA, Venezia, Italy) was used as reference food and served twice to the volunteers. The test was conducted at around 09:00 a.m. The volunteers were given the following instructions: prior to the test they had to fast for 12 h during which they were allowed to drink water *ad libitum*. On the evening before the test, they were allowed to have dinner as usual with the exception of alcohol, which was forbidden. On the test day, blood glucose after fasting was measured at time zero (T0), after which the volunteers consumed pasta or the reference food within a period of 15 min and were given 250 mL of water. Blood glucose was then measured at 15, 30, 45, 60, 90, and 120 min after the volunteers had started to eat. The volunteers were asked to minimize their physical activity during the test. Capillary blood samples were obtained by the finger prick method (Glucoject Dual Plus, A.Menarini diagnostics S.r.l., Firenze, Italy) and glucose was measured using a Glucomen Lx3 calibrated self-monitoring system (A. Menarini diagnostics S.r.l., Firenze, Italy) with a published <3.6% analytical coefficient of variation (CV).

2.6.2. Data Manipulation

The curves of blood glucose increase (μmol ml^{-1}) versus time (from 0 to 120 min) were built for each volunteer after ingestion of the glucose solution and pasta samples (S, WS, SS and SWS). Blood glucose increases were calculated by subtracting the blood glucose concentration value at T0 from the blood glucose concentration values measured at any following time point. In this way, the blood glucose value at T0 was zero for all the curves. The total glucose response (GR) was calculated for each curve as the total area under the curve, by adding the incremental areas under the curve (IAUC) approximated with the trapezoid rule for time frames 0–15, 15–30, 30–45, 45–60, 60–90, and 90–120 min. The height of the rectangular trapezium was the time difference on the x-axis between two subsequent time points (min), whereas the major and minor basis were the values of the blood glucose increases on the y-axis (μmol ml^{-1}), corresponding to the two time points. Finally, the percentage ratio between the GR of the pasta and the GR of the glucose solution, used as reference food, was calculated and hereafter referred to as "apparent glycemic index" (GI_a). The glycemic load (GL) was then calculated using the following Equation (2):

$$GL = (GI_a \times \text{grams of carbohydrate in the standard serving size}/100) \quad (2)$$

2.7. Statistical Analysis

The standard ANOVA procedure (randomized complete design with 2^2 treatments) was applied to the dataset. The experiment involved two factors: with and without the addition of sourdough, and with and without the addition of wholemeal semolina. A multiple comparison procedure was applied for the GR values to determine if the mean values of each sample were significantly different from that of the reference food. The mean values were separated by LSD test at $p = 0.05$ significance level, using the Statgraphics Centurion 18 software package (version 18, Statpont Technologies Inc., Warrenton, VA, USA).

3. Results and Discussion

3.1. In Vitro Starch Digestion

The amount of glucose released from starch during in vitro digestion of the four samples of pasta is reported in Figure 1A,B, while the GIe data are reported in Table 2. The shape of the curves obtained was similar for the four types of pasta and, as expected, the glucose values were far lower than the values obtained from the in vitro digestion of white breadcrumbs reported in the same figure.

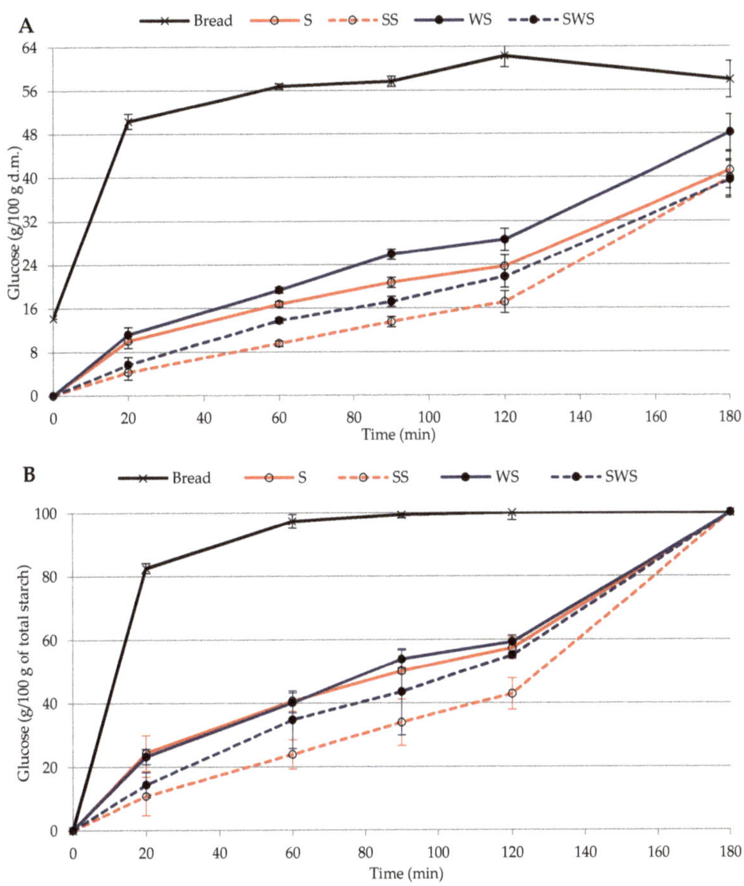

Figure 1. Glucose values from in vitro starch digestion of pasta samples. (**A**) Glucose with respect to dry pasta or bread (g/100 g). (**B**) Glucose with respect to total starch (g/100 g). S, pasta with semolina. SS, pasta with semolina-based sourdough. WS, pasta with wholemeal semolina. SWS, pasta with wholemeal semolina-based sourdough. Bars indicate LSD intervals at 95% confidence level.

Table 2. Results of in vivo digestion and in vitro starch hydrolysis of pasta.

Samples	GR Mg dL^{-1} min^{-1}	GI$_a$	GIe	GL	TS g/100 g Pasta	RDS	SDS g/100 g TS	IDS
S	1584 ^{b}	38.0 ^{b}	33.2 ^{b}	23.9 ^{c}	41.2 ^{b}	24.8 ^{a}	33.0 ^{b}	45.6 ^{b}
SS	1553 ^{b}	41.0 ^{b}	23.6 ^{d}	21.8 ^{d}	39.9 ^{b}	10.7 ^{b}	30.9 ^{b}	56.7 ^{a}
WS	2209 ^{a}	57.0 ^{a}	38.5 ^{a}	31.3 ^{a}	48.1 ^{a}	23.2 ^{a}	35.9 ^{a}	40.8 ^{c}
SWS	2277 ^{a}	55.5 ^{a}	28.6 ^{c}	27.8 ^{b}	39.5 ^{b}	14.4 ^{b}	40.6 ^{a}	45.0 ^{b}
Reference Food	4484	100	100		48.1	82.5	27.4	0
					Significance			
Sourdough	ns	ns	***	*	*	**	ns	*
WholemealSemolina	***	***	***	*	*	ns	*	*
Sourdough*WholemealSemolina	ns	ns	ns	ns	***	ns	ns	ns

*** $p < 0.001$; ** $p < 0.01$; * $p < 0.05$; ns, not significant. The different superscript letters in the column denote a statistically significant difference at $p \leq 0.05$. GR, glucose response. GI$_a$, apparent glycemic index. GIe, glycemic index estimated after in vitro digestion. GL, glycemic load per 160 g serving size. TS, total starch. RDS, rapidly digestible starch. SDS, slowly digestible starch. IDS, inaccessible digestible starch.

In vitro starch hydrolysis of both pasta samples with wholemeal semolina (WS and SWS) released more glucose than the corresponding samples prepared with semolina, indicating that the presence of fiber increased starch availability to the hydrolytic enzymes, in pasta both with and without sourdough. As a consequence, the values of GIe were significantly higher in WS (38.5) than in S (33.2) and in SWS (28.6) than in SS (23.6). This result is in accordance with Vignola et al. [21], who found a higher amount of hydrolyzed starch in wholemeal pasta and postulated that fiber may disrupt protein matrix and give rise to a porous structure that facilitates the action of hydrolytic enzymes. On the contrary, other researchers reported a reduction in starch digestion after the addition of fiber in the pasta formulation. Padalino et al. [22] observed lower starch hydrolysis in wholemeal spaghetti than in semolina spaghetti. A reduction in starch hydrolysis was also observed in pasta made using quinoa flour compared to semolina pasta [23] and the reason was suggested to be due to the high concentration of dietary fiber in quinoa flour. The data reported in Figure 1A show that use of sourdough had a significant effect on the hydrolysis of starch. The hydrolysis curves of pasta with sourdough (SS and SWS) showed lower values of released glucose at any time the analyses were performed, while the GIe was significantly lower than the corresponding pasta made without sourdough (S and WS). This was in accordance with the results obtained by Lorusso et al. [23], who observed a significant decrease in starch hydrolysis in pasta made with fermented quinoa. The lowest values of hydrolyzed starch were found in pasta containing semolina-based sourdough (SS) (Figure 1A). Similar results were observed in sourdough bread, where the degree of in vitro starch digestion was found to be lower than in yeasted bread, with a further decrease in sourdough bread enriched with fibers [24,25], contrary to this work. Reduced starch hydrolysis in pasta with sourdough can be explained as an effect of biological acidification, which creates interactions between gluten and starch, and hinders the access of enzymes to the starch [26].

When the data are expressed as percentage of starch digested over the total (Figure 1B), it is worth mentioning that the rate of hydrolysis in pasta with semolina and pasta with wholemeal semolina is similar. This in agreement with Bustos et al. [27] who reported that the kinetic constant during wholemeal pasta digestion did not differ from that observed in white wheat-based pasta, although starch hydrolysis was higher in wholemeal pasta, as in our data. In the same way, the effect of fermentation is still evident when the data are expressed as percentage of starch digested over the total.

The values of total starch (TS) at 180 min, rapidly digestible starch (RDS), slowly digestible starch (SDS) and inaccessible digestible starch (IDS) were calculated as percentages of TS and are reported in Table 2. Both use of sourdough and use of wholemeal semolina had a significant effect on the values of the starch fractions (Table 2).

The highest value of TS was found in WS pasta (48.1%). It was similar to the value found in white bread and was higher than that of S pasta (41.2%) The value decreased in SWS pasta (39.5%) as an effect of the use of sourdough. Use of sourdough only had an effect on TS in the presence of wholemeal semolina, whereas no differences were detected between S and SS pasta. Demirkesen-Bicak et al. [28] found a lower value of TS in wholemeal bread than in white bread, both in the one fermented with baker's yeast and sourdough, while sourdough fermentation had no effect on TS. As highlighted by Bustos et al. [27], the disintegration kinetics of food directly affects the level of starch hydrolysis, as the more the structure of the food is porous the higher the starch hydrolysis level will be, while the whole wheat bread described by Demirkesen-Bicak et al. [28] had a more compact structure with a lower specific volume than white bread. On the contrary, wholemeal pasta has a more porous structure than semolina pasta due to the presence of insoluble bran fiber, which may disrupt the protein matrix, also leading to a weakening of the structure [27]. The more compact structure of semolina pasta results in a very close protein network, which entraps starch granules and delays α-amylase activity [21]. It has been demonstrated that the presence of bran from wholemeal semolina weakens the pasta structure [29] since it probably interferes with the formation of such a continuous matrix. This hypothesis also explains why the SDS value reported in Table 2 rose from 33.0% in S to 35.9% in WS and from 30.9% in SS to 40.6% in SWS, while the IDS value decreased when wholemeal semolina was used (from 45.6% in S to 40.8% in WS and from 56.7% in SS to 45% in SWS). In fact, it is worth mentioning that IDS is defined as inaccessible digestible starch, which is starch resistant to enzymatic digestion because of the food structure [30]. IDS should include both type 1 and type 3 resistant starch (RS), the former being potentially digestible starch, but physically inaccessible to hydrolytic enzymes, and the second being retrograded starch from food processing [31]. In our case, starch was digested immediately after the thermal treatment of cooking, when the starch was fully gelatinized, so we can infer that almost all the IDS is represented by type 1 resistant starch [16].

The use of sourdough had a significant and positive effect on the IDS content of pasta. In actual fact, an increase in IDS was found when sourdough was used, in both semolina and wholemeal semolina pasta. There is a lack of literature on the use of sourdough in fresh pasta, but data reported by Fois et al. [15] are in line with these data. Demirkesen-Bicak et al. [28] confirmed the increase in resistant starch as an effect of sourdough fermentation, although these data refer to bread, in which case the resistant starch comprises both types 1 and 3.

An explanation of the higher level of IDS in pasta with sourdough can be inferred from the observation of Östman et al. [26], who suggested that the organic acids present in sourdough bread during thermal treatment can promote the formation of starch-gluten interactions, which make the food structure less susceptible to hydrolytic enzymes in the first two hours of digestion. Starch-gluten interactions contribute to type 1 resistant starch. This is confirmed by the lower amount of RDS in pasta with sourdough as compared to pasta without sourdough, in both the semolina and wholemeal semolina samples. Use of sourdough reduced the RDS and increased the IDS levels. These results could have an important implication from a nutritional point of view, as the target outcomes in food processing focus on lowering the RDS and raising the RS content.

3.2. Glycemic Response and Glycemic Load

The curves of glucose increases versus time and the GR values are reported in Figure 2 and Table 2, respectively, for the four pasta samples and for the glucose solution used as a reference food.

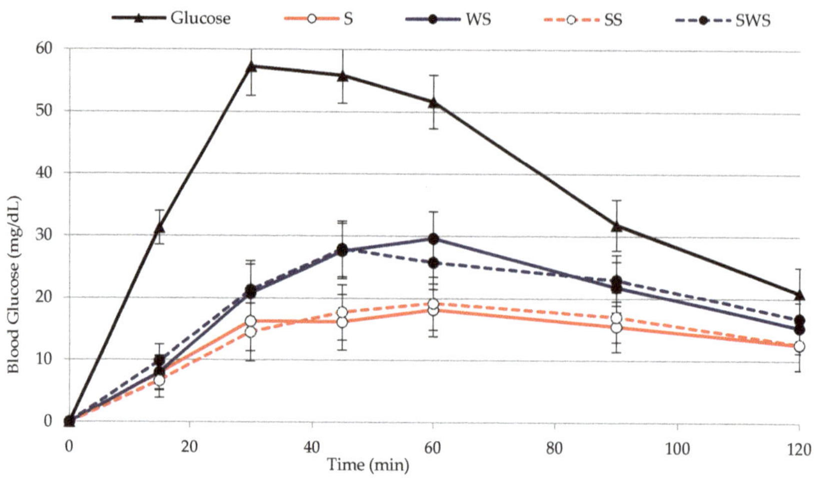

Figure 2. Blood glucose values obtained after in vivo digestion of the four pasta samples. S, pasta with semolina. WS, pasta with wholemeal semolina. SS, pasta with semolina-based sourdough. SWS, pasta with wholemeal semolina-based sourdough. Bars indicate LSD intervals at 95% confidence level.

A comparison of the differences between the mean GR value of each of the pasta samples and the mean GR value of the reference food, showed that such differences were, as expected, highly significant ($p < 0.001$). The data in Table 2 show that use of wholemeal semolina had a significant effect ($p < 0.001$) on glycemic response. On the contrary, sourdough had no effect on glycemic response (Table 2) and the curves corresponding to S and SS, and to WS and SWS were very close (Figure 2). GI_a was calculated as the percentage ratio between the incremental area obtained after ingestion of pasta and the incremental area obtained after ingestion of the reference food, both containing 50 g of available carbohydrates (ACH). The GI_a values were 38.0 and 41.0 for S and SS pasta, and 57.0 and 55.5 for WS and SWS pasta, respectively. Statistical analysis revealed that use of sourdough had no effect on GI_a, whereas use of wholemeal semolina did (Table 2), and that there was no interaction between the two factors (wholemeal semolina and sourdough). Foods are commonly divided into three classes on the basis of their GI, i.e., foods with low GI (<55); foods with intermediate GI (55–70); foods with high GI (>70). According to this classification, the values obtained here indicate that the addition of wholemeal semolina to the pasta formulation raised the GI_a of the pasta from low to medium.

The effect of the use of wholemeal semolina was unexpected. Henry et al. [32] reported that there was no difference between the GI values of Fusilli pasta and whole wheat Fusilli pasta. Atkinson et al. [33] reported that the average GI of white spaghetti (49) and wholemeal spaghetti (48), derived from multiple studies by different laboratories, was the same. Kristensen et al. [34] reported that fiber had no effect on postprandial glycemia, when comparing refined wheat pasta and whole wheat pasta-based meals. Those authors hypothesized that this lack of fiber having an effect might be related to the type of dietary fiber present in wheat which, as it does not form a viscous solution upon hydration in the gastrointestinal tract, probably does not delay gastric emptying. In this work, the higher glycemic response of wholemeal-based pasta observed in vivo was in agreement with the data obtained after in vitro starch hydrolysis (Figure 1), which suggests that the fiber favored the accessibility of hydrolytic enzymes to starch granules, both in vitro and in vivo. It should be noted that no data can be found on the glycemic response of a pasta such as the one analyzed in this work (i.e., fresh and pasteurized gnocchetti-type pasta), and that all available data in literature are on dry pasta, mainly the spaghetti type. Scazzina et al. [35] published the GI of certain commercial Italian foods, among which pasta, showing how the GI of wholemeal spaghetti can range from low (35) to almost medium (55) depending on the brand. Meaning that different processing conditions can drastically modify the GI value.

In this study, use of sourdough had no effect on the glycemic response of fresh pasta, contrary to what was found in literature for bread, which reported how sourdough fermentation or the addition of organic acids had been used to lower the glycemic index [36]. This, as organic acids were thought to delay gastric emptying (mainly acetic acid), or to promote starch-gluten interactions which reduce starch bioavailability (mainly lactic acid) after heat treatment [26]. Scazzina et al. [37] found, for bread, that neither the leavening technique nor fiber content influenced starch availability to hydrolytic enzymes in vitro, and that the leavening technique significantly affected glucose response in vivo, whereas fiber content did not, suggesting that organic acids could delay gastric emptying without influencing starch availability. A possible explanation of our in vivo data cannot be related to the loss of organic acids in the cooking water, as reported by Fois et al. [11], although the pasta was cooked in water before in vitro digestion, where an effect of fermentation was detected. The pasta for in vivo measurements was cooked in salted water to make it palatable, whereas the pasta for in vitro measurement was not. A possible effect of salt on the gluten network and on starch gelatinization might have interfered with the effect of sourdough fermentation and needs further investigation.

It is worth mentioning that the increase in postprandial glycemia is not only related to the GI of a food but also to the amount of carbohydrates in the serving size of that food. Thus, another index, glycemic load (GL), was used [36]. In this study GL was calculated as the product of GI_a and the grams of available carbohydrate in 160 g of cooked pasta (standard serving size) divided by 100. As reported in Table 1, the amount of ACH in cooked semolina-based pasta (39.24% in S and 33.28% in SS) was higher ($p < 0.05$) than that of wholemeal-based pasta (34.27% in WS and 31.32% in SWS). Then, the serving size of pasta containing 50 g of ACH consumed by the volunteers, was different (i.e., 127 g of S and 150 g of SS were weighed), whereas in the case of WS and SWS the weight was 146 and 159 g, respectively. This result was in accordance with a study by Henry et al. [32], who reported a lower content of available carbohydrates and, therefore, a larger serving size for whole wheat Fusilli than semolina Fusilli. Moreover, even sourdough wholemeal bread was found to have a lower content of available carbohydrates than yeast leavened bread [38]. The GL values were 23.9 for S, 21.8 for SS, 31.3 for WS, and 27.8 for SWS. The effect of using wholemeal semolina was still evident in GL, but a new effect of the use of sourdough, which reduced the GL value, could be detected giving pasta with sourdough an increased value from a nutritional point of view.

4. Conclusions

Pasta is a popular carbohydrate-based food with a low glycemic index, which can be fortified with a variety of ingredients to improve its nutritional qualities. In this paper, the digestibility of starch was studied, in vitro and in vivo, in pasta with the addition of wholemeal semolina and sourdough. The results showed that use of wholemeal semolina made the gluten matrix more susceptible to the in vitro activity of hydrolytic enzymes. This result was confirmed by glucose determination after in vivo digestion; unexpectedly the GI_a value of wholemeal pasta was higher than that of semolina pasta. However, the data available in literature concern dry pasta, mostly the spaghetti type, whereas gnocchetti-type fresh and pasteurized pasta was prepared for this study. It is known that food processing conditions drastically affect the activity of hydrolytic enzymes and, as a consequence, the GI of foods. Use of sourdough had no effect on the GR value, whereas the GL value decreased owing to the effect of the reduced content of available carbohydrates in pasta made with sourdough. It is rather difficult to compare these results with the data available in literature, which is focused on the use of sourdough in bread making. The physical characteristics of the matrices are deeply different. In the case of bread, the crumb is a porous matrix with a high surface to volume ratio. It is therefore more susceptive to the attack of hydrolytic enzymes than pasta which, as a consequence of the extrusion process, has a more compact structure and a reduced surface to volume ratio. Moreover, pasta is cooked in an excess of hot water, in which some of the starch is leached. Finally, pasta is eaten immediately after cooking, without leaving time for the retrogradation phenomenon to fully take place.

Author Contributions: P.C. conceived, designed and coordinated the study, drafted and revised the manuscript. S.F. conducted the study, performed the statistical analysis and the interpretation of results and drafted the manuscript. P.P.P. and M.S. prepared the sourdough, made the pasta samples, and performed the in vitro digestion of pasta. T.R. was responsible for managing the funds. All authors have read and agreed to the published version of the manuscript.

Funding: This research was funded by Sardegna Ricerche, Science and Technology Park of Sardinia, and Regione Autonoma della Sardegna, under the program POR FESR 2014/2020-Asseprioritario I-Ricerca scientifica, sviluppo tecnologico e innovazione. Progetto cluster top down PAFASEF.

Institutional Review Board Statement: The study was conducted according to the guidelines of the Declaration of Helsinki, and approved by the Ethics Committee of ATS Sardegna, protocol code 118/2018, date of approval 04.012.2018.

Informed Consent Statement: Informed consent was obtained from all subjects involved in the study.

Data Availability Statement: The data presented in this study are available on request from the corresponding author.

Acknowledgments: The authors want to thank the volunteers who kindly participated to the IG study for their stoicism and their contribution to the science. Acknowledgement goes to Martina Serra for her help in conducting the IG study, and to Sergio Uzzau for his medical assistance.

Conflicts of Interest: The authors declare that there is no conflict of interest in this research article. The authors are researchers employed by Porto Conte Ricerche Srl, a publicly-owned research company, subsidiary controlled and financed by Sardegna Ricerche, which is an agency of the Regional government of Sardinia.

References

1. IPO International Pasta Organisation. Available online: https://internationalpasta.org/news/ (accessed on 14 June 2021).
2. Giacco, R.; Vitale, M.; Riccardi, G. Pasta: Role in Diet. In *Encyclopedia of Food and Health*; Elsevier BV: Berlin, Germany, 2016; Volume 4, pp. 242–245.
3. Rawat, N.; Indrani, D. Functional Ingredients of Wheat-Based Bakery, Traditional, Pasta, and Other Food Products. *Food Rev. Int.* **2014**, *31*, 125–146. [CrossRef]
4. Aune, D.; Chan, D.S.M.; Lau, R.; Vieira, R.; Greenwood, D.C.; Kampman, E.; Norat, T. Dietary fibre, whole grains, and risk of colorectal cancer: Systematic review and dose-response meta-analysis of prospective studies. *BMJ* **2011**, *343*, d6617. [CrossRef]
5. Hirawan, R.; Beta, T. Whole Wheat Pasta and Health. In *Wheat and Rice in Disease Prevention and Health*; Elsevier BV: Berlin, Germany, 2014; pp. 5–16.
6. Cioffi, I.; Ibrugger, S.; Bache, J.; Thomassen, M.T.; Contaldo, F.; Pasanisi, F.; Kristensen, M. Effects on satiation, satiety and food intake of wholegrain and refined grain pasta. *Appetite* **2016**, *107*, 152–158. [CrossRef]
7. Rackerby, B.; Kim, H.J.; Dallas, D.C.; Park, S.H. Understanding the effects of dietary components on the gut microbiome and human health. *Food Sci. Biotechnol.* **2020**, *29*, 1463–1474. [CrossRef]
8. Scazzina, F.; Siebenhandl-Ehn, S.; Pellegrini, N. The effect of dietary fibre on reducing the glycaemic index of bread. *Br. J. Nutr.* **2013**, *109*, 1163–1174. [CrossRef]
9. Hemdane, S.; Jacobs, P.J.; Dornez, E.; Verspreet, J.; Delcour, J.; Courtin, C.M. Wheat (*Triticum aestivum* L.) Bran in Bread Making: A Critical Review. *Compr. Rev. Food Sci. Food Saf.* **2015**, *15*, 28–42. [CrossRef]
10. Poutanen, K.; Flander, L.; Katina, K. Sourdough and cereal fermentation in a nutritional perspective. *Food Microbiol.* **2009**, *26*, 693–699. [CrossRef] [PubMed]
11. Fois, S.; Campus, M.; Piu, P.P.; Siliani, S.; Sanna, M.; Roggio, T.; Catzeddu, P. Fresh Pasta Manufactured with Fermented Whole Wheat Semolina: Physicochemical, Sensorial, and Nutritional Properties. *Foods* **2019**, *8*, 422. [CrossRef] [PubMed]
12. Vitale, M.; Masulli, M.; Rivellese, A.A.; Bonora, E.; Babini, A.C.; Sartore, G.; Corsi, L.; Buzzetti, R.; Citro, G.; Baldassarre, M.P.A.; et al. Pasta Consumption and Connected Dietary Habits: Associations with Glucose Control, Adiposity Measures, and Cardiovascular Risk Factors in People with Type 2 Diabetes—TOSCA.IT Study. *Nutrients* **2019**, *12*, 101. [CrossRef] [PubMed]

13. Chiavaroli, L.; Kendall, C.W.C.; Braunstein, C.R.; Mejia, S.B.; Leiter, A.L.; Jenkins, A.D.J.; Sievenpiper, J.L. Effect of pasta in the context of low-glycaemic index dietary patterns on body weight and markers of adiposity: A systematic review and meta-analysis of randomised controlled trials in adults. *BMJ Open* **2018**, *8*, e019438. [CrossRef]
14. Livesey, G.; Taylor, R.; Hulshof, R.; Howlett, J. Glycemic response and health—A systematic review and meta-analysis: Re-lations between dietary glycemic properties and health outcomes. *Am. J. Clin. Nutr.* **2008**, *87*, 258S–268S. [CrossRef] [PubMed]
15. Fois, S.; Piu, P.P.; Sanna, M.; Roggio, T.; Catzeddu, P. Starch digestibility and properties of fresh pasta made with semoli-na-based liquid sourdough. *LWT-Food Sci. Technol.* **2018**, *89*, 496–502. [CrossRef]
16. Sanna, M.; Fois, S.; Falchi, G.; Campus, M.; Roggio, T.; Catzeddu, P. Effect of liquid sourdough technology on the pre-biotic, texture, and sensory properties of a crispy flatbread. *Food Sci. Biotechnol.* **2019**, *28*, 721–730. [CrossRef]
17. Ficco, D.B.M.; De Simone, V.; De Leonardis, A.M.; Giovanniello, V.; Del Nobile, M.A.; Padalino, L.; Lecce, L.; Borrelli, G.M.; De Vita, P. Use of purple durum wheat to produce naturally functional fresh and dry pasta. *Food Chem.* **2016**, *205*, 187–195. [CrossRef]
18. Food and Agriculture Organization/World Health Organization. Carbohydrates in Human Nutrition. Report of a Joint FAO/WHO Expert Consultation. *FAO Food Nutr. Pap.* **1998**, *66*. Available online: http://www.fao.org/3/W8079E/w8079e00.htm#Contents (accessed on 18 October 2021).
19. Sugiyama, M.M.; Tang, A.C.; Wakaki, Y.; Koyama, W. Glycemic index of single and mixed meal foods among common Japanese foods with white rice as a reference food. *Eur. J. Cli. Nutr.* **2003**, *57*, 743–752. [CrossRef]
20. *AACC Approved Methods of Analysis*, 11th ed; Cereals & Grains Association: St. Paul, MN, USA, Method 66-50.01; Available online: http://methods.aaccnet.org/toc.aspx (accessed on 18 October 2021).
21. Vignola, M.B.; Bustos, M.C.; Pérez, G.T. In vitro dialyzability of essential minerals from white and whole grain pasta. *Food Chem.* **2018**, *265*, 128–134. [CrossRef]
22. Padalino, L.; Mastromatteo, M.; Lecce, L.; Spinelli, S.; Conte, A.; Del Nobile, M.A. Effect of raw material on cooking quality and nutritional composition of durum wheat spaghetti. *Int. J. Food Sci. Nutr.* **2015**, *66*, 266–274. [CrossRef]
23. Lorusso, A.; Verni, M.; Montemurro, M.; Coda, R.; Gobbetti, M.; Rizzello, C.G. Use of fermented quinoa flour for pasta making and evaluation of the technological and nutritional features. *LWT* **2017**, *78*, 215–221. [CrossRef]
24. De Angelis, M.; Damiano, N.; Rizzello, C.G.; Cassone, A.; Di Cagno, R.; Gobbetti, M. Sourdough fermentation as a tool for the manufacture of low-glycemic index white wheat bread enriched in dietary fibre. *Eur. Food Res. Technol.* **2009**, *229*, 593–601. [CrossRef]
25. Rizzello, C.G.; Lorusso, A.; Montemurro, M.; Gobbetti, M. Use of sourdough made with quinoa (*Chenopodium quinoa*) flour and autochthonous selected lactic acid bacteria for enhancing the nutritional, textural and sensory features of white bread. *Food Microbiol.* **2016**, *56*, 1–13. [CrossRef]
26. Östman, E.; Nilsson, M.; Elmståhl, H.L.; Molin, G.; Björck, I. On the Effect of Lactic Acid on Blood Glucose and Insulin Responses to Cereal Products: Mechanistic Studies in Healthy Subjects and In Vitro. *J. Cereal Sci.* **2002**, *36*, 339–346. [CrossRef]
27. Bustos, M.C.; Vignola, M.B.; Perez, G.; León, A.E. In vitro digestion kinetics and bioaccessibility of starch in cereal food products. *J. Cereal Sci.* **2017**, *77*, 243–250. [CrossRef]
28. Demirkesen-Bicak, H.; Arici, M.; Yaman, M.; Karasu, S.; Sagdic, O. Effect of Different Fermentation Condition on Estimated Glycemic Index, In Vitro Starch Digestibility, and Textural and Sensory Properties of Sourdough Bread. *Foods* **2021**, *10*, 514. [CrossRef] [PubMed]
29. Chillo, S.; Laverse, J.; Falcone, P.M.; Protopapa, A.; Del Nobile, M. Influence of the addition of buckwheat flour and durum wheat bran on spaghetti quality. *J. Cereal Sci.* **2008**, *47*, 144–152. [CrossRef]
30. Mishra, S.; Monro, J.A. Digestibility of starch fractions in whole grain rolled oats. *J. Cereal Sci.* **2009**, *50*, 61–66. [CrossRef]
31. Fuentes-Zaragoza, E.; Riquelme-Navarrete, M.J.; Sànchez-Zapata, E.; Pérez-Alvarez, J.A. Resistant starch as functional in-gredient: A review. *Food Res. Int.* **2010**, *43*, 931–942. [CrossRef]
32. Henry, C.J.K.; Lightowler, H.J.; Strik, C.M.; Renton, H.; Hails, S. Glycaemic index and glycaemic load values of commer-cially available products in the UK. *Brit. J. Nutr.* **2005**, *94*, 922–930. [CrossRef]
33. Atkinson, F.S.; Foster-Powell, K.; Brand-Miller, J.C. International tables of glycemic index and glycemic load values. *Diabetes Care* **2008**, *31*, 2281–2283. [CrossRef]
34. Kristensen, M.; Jensen, M.G.; Riboldi, G.; Petronio, M.; Bügel, S.; Toubro, S.; Tetens, I.; Astrup, A. Wholegrain vs. refined wheat bread and pasta. Effect on postprandial glycemia, appetite, and subsequent ad libitum energy intake in young healthy adults. *Appetite* **2010**, *54*, 163–169. [CrossRef]
35. Scazzina, F.; Dall'Asta, M.; Casiraghi, M.C.; Sieri, S.; Del Rio, D.; Pellegrini, N.; Brighenti, F. Glycemic index and glycemic load of commercial Italian foods. *Nutr. Metab. Cardiovasc. Dis.* **2016**, *26*, 419–429. [CrossRef] [PubMed]
36. Björck, I.; Liljeberg Elmståhl, H. The glycemic index: Importance of dietary fibre and other food properties. *Porc. Nutr. Soc.* **2003**, *62*, 201–206. [CrossRef] [PubMed]
37. Scazzina, F.; Del Rio, D.; Pellegrini, N.; Brighenti, F. Sourdough bread: Starch digestibility and postprandial glycemic response. *J. Cereal Sci.* **2009**, *49*, 419–421. [CrossRef]
38. Lappi, J.; Selinheimo, E.; Schwab, U.; Katina, K.; Lehtinen, P.; Mykkanen, H.; Kolehmainen, M.; Poutanen, K. Sourdough fermentation of wholemeal wheat bread increases solubility of arabinoxylan and protein and decreases postprandial glucose and insulin responses. *J. Cereal Sci.* **2010**, *51*, 152–158. [CrossRef]

foods

MDPI

Article

Gluten-Free Pasta Enriched with Fish By-Product for Special Dietary Uses: Technological Quality and Sensory Properties

Andrea Aínsa, Alba Vega, Adrian Honrado, Pedro Marquina, Pedro Roncales, José A. Beltrán Gracia and Juan B. Calanche Morales *

Instituto Agroalimentario de Aragón -IA2, Universidad de Zaragoza-CITA, Miguel Servet. 177, 50013 Zaragoza, Spain; andreaainsa6@gmail.com (A.A.); vega.glez.alba@gmail.com (A.V.); 742755@unizar.es (A.H.); pmarquin@gmail.com (P.M.); roncales@unizar.es (P.R.); jbeltran@unizar.es (J.A.B.G.)

* Correspondence: calanche@unizar.es

Abstract: Gluten-free pasta enriched with fish can support a nutritive and suitable option for people with celiac disease that allows achieving the benefits of fish consumption, especially the consumption of Ω-3 fatty acids; however, this requires that the pasta has adequate technological and sensory properties. For this purpose, four optimal formulations, obtained with an iterative process, were analyzed to determine the effect of the different ingredients (yellow corn flour, white corn flour, and rice flour) in gluten-free pasta compared to commercial wheat pasta. An evaluation of the color, texture, and technological properties were conducted, and the pasta was sensorially characterized. The enriched gluten-free pasta required shorter cooking times (≈3 min) and was characterized by lower hardness, springiness, gumminess, chewiness, and fracturability, and had higher values of adhesiveness than wheat pasta. In addition, the incorporation of yellow corn gives gluten-free pasta a similarity in color to commercial pasta, with a value of ΔE between 5.5 and 8.0. Regarding the sensory analysis, gluten-free pasta was characterized by slight fishy aromas and flavors with some aftertaste compared to commercial pasta. Finally, the use of different cereals to obtain gluten-free pasta could be a good and feasible alternative despite the technological and sensory modifications observed.

Keywords: texture profile analysis; gluten-free; celiac disease; pasta; fish; allergy

Citation: Aínsa, A.; Vega, A.; Honrado, A.; Marquina, P.; Roncales, P.; Gracia, J.A.B.; Morales, J.B.C. Gluten-Free Pasta Enriched with Fish by-Product for Special Dietary Uses: Technological Quality and Sensory Properties. *Foods* **2021**, *10*, 3049. https://doi.org/10.3390/foods10123049

Academic Editor: Talwinder S. Kahlon

Received: 6 November 2021
Accepted: 7 December 2021
Published: 8 December 2021

1. Introduction

Pasta, in Spanish legislation, is a term employed to describe those products obtained by desiccation of an unfermented dough made with semolina or flour derived from durum wheat, semi-hard wheat, soft wheat, or a mixture of these, and drinking water [1]. Moreover, this product is highly consumed due to its ease of preparation, great versatility, low cost, and good organoleptic properties.

However, this type of pasta cannot be consumed by a part of the population due to celiac disease, which is defined as an autoimmune and multisystemic enteropathy, caused by gluten and prolamins that affect subjects who are genetically predisposed [2,3]. For this reason, people look for gluten-free food with a similar appearance to conventional products, particularly food options that are organoleptically acceptable, with great nutritional features, and are economically accessible. Cereals that affect celiac disease and, therefore, should be removed from the diet are wheat (*Triticum aestivum*), rye (*Secale cereale*), triticale (*Triticum* spp x *Secale cereale*), barley (*Hordeum vulgare*), kamut (*Triticum turqidum*), spelt (*Triticum spelta*), and varieties of oats (*Avena sativa*) that are not guaranteed to be free from cross-contamination and are not certified as gluten-free. Gluten-free foods can be, on the one hand, generics that do not contain gluten by nature and, on the other hand, non-generics that are divided into conventional ones: those whose formulation or preparation may contain traces of gluten due to cross-contamination, and specific: products specially formulated for people with celiac disease [4]. Gluten-free pasta products are foods that have been specially manufactured, elaborated, and processed to replace gluten and whose gluten level does not exceed 20 ppm of gluten, relative to the product that reaches the final consumer, as indicated in Reglamento 828/2014 [5].

In addition, most gluten-free products, in general, have shown poorer nutritional quality with a low value of minerals and protein with respect to wheat pasta [6]. Further, the elimination of wheat in gluten-free products entails major modifications in nutritional parameters and sensory quality such as color, flavor, and texture [7]. However, due to the possibility they offer, the use of cereals and pseudo-cereals, which are allowed as ingredients in this type of specific formulation, continues to be promoted, since in most cases, it creates a notable improvement of the nutritional properties. To solve the technological difficulties, hydrocolloids, emulsifiers, dairy derivatives, eggs, or soy protein are used in order to improve the texture [8]. According to other authors, an effective alternative strategy to improve nutritional composition and rheological properties of pasta is adding distinct ingredients to the dough, for example, fish [9–13]. Fish is an excellent source of proteins, lipids rich in unsaturated fatty acids, especially Ω3 such as eicosapentaenoic acid (EPA) and docosahexaenoic acid (DHA), minerals, and vitamins (A, D, B6, and B12) [14]. Regarding the above subject, the gluten-free pasta enriched with Mechanically Deboned Meat (MDM) from seabass (*Dicentrachus labrax*) by-products previously dried and milled, represents a new alternative product that could encourage fish consumption from people who do not eat it frequently, which would also help to prevent cardiovascular diseases due to the enriching with bioactive compounds such as unsaturated fatty acids, especially of Ω-3 type, which can remain stable without reaching a sensory rejection as has been seen in previous studies due to the use of an antioxidant [9,11,15]. On top of that, gluten-free pasta would allow taking advantage of the waste of fish processing (offcutting); at the same time, it could be an opportunity to become more environmentally friendly, and thus, allowing for more sustainable activity [16].

However, the change of original ingredients for different ones in pasta could affect physicochemical, sensory, and technological properties. The quality and characteristics of cooked pasta are determined by different parameters such as optimal cooking time, weight gain, hydration, losses during cooking, and texture measured by instrument and sensory analysis [17]. For that reason, gluten-free pasta production focuses on maintaining quality and sensory parameters similar to conventional durum wheat pasta, which could improve its acceptability by consumers.

This research had a focus on the evaluation of the physical, technological, and sensory properties of gluten-free pasta enriched with bioactive compounds from seabass by-products that were added to offer adapted to the celiac population that could contribute to a healthy diet. The main purpose was to obtain a gluten-free pasta with adequate quality and a similarity to traditional wheat pasta.

2. Materials and Methods

2.1. Raw Material

Fish used to elaborate gluten-free pasta was an offcutting of seabass (*Dicentrachus labrax*) from the filleting process and were provided by a local fish industry (Scanfisk®, Zaragoza, Spain). Bones and skin were removed, and the fillets were dipped in saline solution 8%. Then, they were dried in an oven at 60 °C for 24 h with slow air velocity and pulverized to obtain the concentrates. Regarding the cereal, different flours bought in the supermarket were chosen. Gums (xanthan gum and locust bean gum) (SOC Chef 8, DELITÉ 9, Gilca) were used to improve the texture. To choose the definitive gluten-free pasta enriched with seabass concentrate, different formulations were made following the methodology described in Calanche et al. (2019) and Ainsa et al. (2021) [9,11], optimized and adapted to obtain optimal formulations, which are shown in Table 1. Due to the fact that the moisture in the different flours (≈10%) was not significant, the water content was that established for a common pasta. Different gluten-free pastas were compared with a commercial durum wheat pasta, which is denominated as "Control", and was provided by a local pasta factory (Pastas Romero®, Daroca, Spain); the Control pasta was used to analyze the physical, technological, and sensory parameters. Innovative pasta was manufactured with an experimental extrusion machine (Bottene, Mod. Lillodue 14057CE, Marano Vicentino, Italy) [15].

Table 1. Formulations of gluten-free pasta with fish concentrate.

Pasta Formulation	Dry Matter Ingredients (75%)	Wet Matter Ingredients (25%)
CONTROL	Durum wheat semolina	Drinking water
YCRO	Yellow corn flour (45%), rice flour (40%), oat bran (5%), seabass concentrate (10%)	Drinking water + gums (0.6%)
YCR	Yellow corn flour (45%), rice flour (45%), seabass concentrate (10%)	Drinking water + gums (0.6%)
WCRO	White corn flour (45%), rice flour (40%), oat bran (5%), seabass concentrate (10%)	Drinking water + gums (0.6%)
ROG	Rice flour (80%), oat bran (10%), seabass concentrate (10%)	Drinking water + gums (6%)

2.2. *Physical Properties*

2.2.1. Estimation of Optimal Cooking Time

The analysis was performed by the visual method following the AACC method 66–50 [18] and by the instrumental method of the Warner–Bratzler shear test for which a rheometer was used (ANAME Scientific Instrumentation, mod. TA-XT2i, Madrid, Spain). The test was performed with the following parameters: pre-test speed: 2 mm/s; test speed: 2 mm/s; post-test speed: 10 mm/s; cutting distance: 15 mm; force threshold: 10 g. After cooking, the pasta was left to cool on a damp paper to avoid drying until it reached 22 °C. The determination was carried out, making a total of 10 consecutive measurements, with a flat probe, to determine, on the one hand, the firmness expressed in kg, described as the maximum force to cut the sample and, on the other hand, the cutting effort, expressed in kg·s.

2.2.2. Texture Profile Analysis (TPA) of Gluten-Free Pasta

The texture profile analysis was made with a texturometer (ANAME Scientific Instrumentation, mod. TA-XT2i, Spain) with a flat cylindrical aluminum and consisted of the application of two compression cycles with a rest time between both (decompression) of 20 s, which allowed the determination of different texture properties: hardness, adhesiveness, cohesiveness, springiness, gumminess, chewiness, resilience, and fracturability. The following conditions were established: test speed: 2 mm/s; sample deformation: 75%; force threshold: 10 g.

2.2.3. Pasta Color

Color readings were taken from three separate points on the surface of cooked pasta (after the pasta was cooked to optimal cooking time, drained, and allowed to stand for 5 min at room temperature). Color measures were made using a colorimeter (Minolta, CM-2002, Osaka, Japan). For each sample, readings were taken 10 times, and the mean value was reported. The following color parameters were recorded: L^* (brightness), a^* (redness) and b^* (yellowness). The color variation produced by the addition of different ingredients was calculated with the total color difference (ΔE), from a commercial durum wheat pasta as a reference, using the following formula:

$$\Delta E = \sqrt{(\Delta L^*)^2 + (\Delta a^*)^2 + (\Delta b^*)^2}$$

where: $\Delta L = L^*$ Fish pasta– L^* Standard pasta; $\Delta a =$ a Fish pasta– a^* Standard pasta; $\Delta b = b^*$ Fish pasta– b^* Standard pasta.

2.3. Technological Properties

2.3.1. Weight Gain and Hydration

The weight gain and hydration of pasta were determined according to the proceedings described by Cleary and Brennan (2006) [19] with the following modifications: 3 g of pasta was cooked in 180 mL of distilled water during the optimal cooking time estimated for each pasta and was cooled in 100 mL of cold water; then, the pasta was dried with absorbent paper and weighed in an analytical balance. To obtain the percentage of the total weight gain of cooked pasta, the following equation was applied:

$$\text{Weight gain} = \frac{\text{Cooked pasta weight} - \text{Raw pasta weight}}{\text{Raw pasta weight}} \times 100 \tag{2}$$

Once the cooked pasta was obtained, it was dehydrated in an oven at 105 °C for 24 h until reaching a constant weight. The swelling index during cooking was calculated by the following formula:

$$\text{Swelling index} = \frac{\text{Cooked pasta weight (g)}}{\text{Dried pasta weight (g)}} \tag{3}$$

2.3.2. Cooking Losses

The AACC 66–50 method was carried out with the same characteristics as in the previous section (3 g of pasta in 180 mL of water) in their respective optimal cooking times previously estimated [18]. The water resulting from the cooking was collected in crucibles and allowed to evaporate on a stove at 105 °C until reaching a constant weight (24 h). The dry residue was weighed on an analytical balance and determined as a percentage of the total weight of the pasta before cooking.

2.3.3. Moisture

The samples were weighed on the analytical balance; first, they were milled with a laboratory mortar; they were left to dry in an oven at 105 °C for 24 h; they were cooled to room temperature in a desiccator for 1 h; finally, they were weighed again on an analytical balance.

$$\text{Moisture (\%)} = \frac{\text{Raw pasta weight} - \text{Dried pasta weight}}{\text{Raw pasta weight}} \times 100 \tag{4}$$

2.4. Sensorial Analysis

2.4.1. Sensory Texture Profile (STP)

Sensory texture profile was established following the procedure described in ISO 11036:2020 [20], with adaptations that were made by a panel of expert sensory assessors [21]. Samples were evaluated by a trained panel containing 10 selected assessors [22] who analyzed the parameters: elasticity (assessed with the hands), hardness (when biting with the incisors), disintegration, graininess, pastiness, and stickiness; in a structured scale from 0 to 5. Where 0 represents the absence of the attribute and 5 its maximum intensity.

2.4.2. Quantitative Descriptive Analysis (QDA)

The panel consisted of 10 selected assessors who analyzed different samples according to QDA methodology purpose by Calanche et al. (2019) and Ainsa (2019) for pasta with fish added [9,11]. Attributes evaluated were characteristic aroma of cooked pasta, fish aroma, other odors, hardness (when biting with the incisors), characteristic flavor of cooked pasta, fish flavor, aftertaste (once the sample disappeared from the mouth), other flavors, characteristic color of pasta (intensity in yellow), and the homogeneity of the color. In the sessions, pasta was prepared by cooking in boiling water (100 °C). The samples were served without any type of accompaniment at a temperature of 60 °C following the recommendations of UNE-ISO 6658:2019 [23].

2.5. Statistical Analysis

Data obtained in this research were processed by descriptive and inferential statistics using XLSTAT software, Version 2016 (Addinsoft©, Paris, France). A univariate analysis was carried out for each considered variable. We conducted a study of distribution to check the normality of data and to detect outlier's values. Then, Analysis of Variance (ANOVA) tests were applied with a 95% confidence interval and followed by a post hoc test (Fisher) to establish significant differences for different assayed treatments. Moreover, Principal Component Analysis (PCA) was performed to obtain an overview of results obtained by TPA and STP, which allowed us to understand the relationship among parameters taking into account the kind of developed gluten-free pasta. Furthermore, sensory data collected from a selected assessor's panel when QDA was carried out were used to obtain specific profiles for pasta developed based on Square Cosines Method (sensory characterization).

3. Results

3.1. Formulations

As shown in Table 1, the durum wheat semolina was replaced in the gluten-free formulations developed in this study by flours of other cereals such as corn (*Zea mays*)—yellow and white—white rice (*Oryza sativa*), and oat (*Avena sativa*) bran. The key aspect to selecting corn was its versatility of use and color, while rice represented an alternative source of starch and protein. It is well known that a decrease in the protein and an increase in the starch content due to the change of the base cereal in the manufacturing of pasta could affect final products that result in a high glycemic index (GI) [24]. Due to the above, increasing the amount of dietary fiber in the pasta through the inclusion of oat bran could, in most of the formulations, reduce GI because it partially replaces the amount of cereal flours.

3.2. Physical Properties

3.2.1. Texture Properties

Table 2 shows the optimal cooking time for gluten-free pasta and the results for TPA; concerning ideal cooking time, it was measured instrumentally, using the Warner–Bratzler cutting test (WB) (with a standard deviation of 0.689), and by visual determination (OCT) with a positive correlation between both methods (r^2 = 0.999).

Table 2. Texture properties at the optimal cooking time.

PASTA	WB	OCT (s)	HARD	ADH	SPRING	COH	GUM	CHEW	FRACT
CONTROL	-	-	3726.35 c	−16.01 d	0.77 b	0.67 c	2545.54 c	1993.26 b	758.42 c
YCRO	0.459 a	200	2116.91 b	−25.41 d	0.56 a	0.54 b	1144.35 b	651.48 a	519.52 ab
YCR	2.762 b	210	2208.85 b	−90.93 b	0.58 a	0.43 a	968.00 ab	564.53 a	583.56 b
WCRO	0.408 a	200	1795.78 a	−42.88 c	0.56 a	0.51 b	919.55 ab	517.76 a	461.26 a
ROG	0.396 a	200	1946.21 ab	−257.83 a	0.54 a	0.45 a	896.92 a	488.92 a	496.06 a

Parameters: WB: Warner–Bratzler; OCT: optimal cooking time; HARD: hardness; AHD: adhesiveness; SPRING: springiness; COH: cohesiveness; GUM: gumminess; CHEW: chewiness; FRACT: fracturability. Treatments: YCRO: Yellow corn rice oat; YCR: Yellow corn rice; WCRO: White corn rice oat; ROG: Rice oat gums. Different letters in the same column indicate significant differences ($p \leq 0.05$) among treatments.

The estimation of optimal cooking time was very similar using both methods, even coinciding in the YCR pasta, which presented significant differences ($p < 0.05$) concerning the rest of the pasta that could be due to the lack of oat bran in its formulation. The cooking time in the commercial durum wheat pasta used as a control was 10 min; in contrast, gluten-free pasta enriched with seabass concentrate presented optimal cooking times around 3 min.

Regarding TPA, in general, parameters correlated very well with each other ($r^2 \geq 0.90$) and were significantly ($p < 0.05$) different from the types of pasta studied.

According to the results, the control pasta was harder than gluten-free pasta, highlighting WCRO pasta with the lowest value. In relation to adhesiveness, in general, gluten-free pasta was significantly higher ($p < 0.05$) than the control pasta, although the control pasta was statistically similar to YCRO. Concerning springiness, gluten-free pasta was significantly ($p < 0.05$) less elastic than the durum wheat control. Cohesiveness was similar between YCRO and WCRO as well as between YCR and ROG, while the control was significantly different ($p < 0.05$) to all gluten-free pasta. In addition, the gumminess of the control pasta was significantly ($p < 0.05$) bigger than the rest of the gluten-free pasta. About chewiness, control pasta showed significant differences ($p < 0.05$) from gluten-free pasta (+ or −). Finally, control pasta was greater ($p < 0.05$) at fracturability than the rest of the gluten-free pasta.

3.2.2. Color

The color parameters (CIE$L^*a^*b^*$ coordinates and ΔE) of gluten-free and control pasta are shown in Table 3.

Table 3. Color parameters for control and gluten-free pasta.

PASTA	L^* (D65)	a^* (D65)	b^* (D65)	ΔE
CONTROL	60.640 bc	−1.608 cd	19.002 c	0.000 d
YCRO	62.369 b	1.074 a	25.769 a	7.788 b
YCR	60.436 c	−1.073 b	21.454 b	5.593 c
WCRO	68.038 a	−2.039 d	8.957 d	12.685 a
ROG	61.408 bc	−1.350 bc	9.893 d	9.332 b

L^*: brightness, a^*: redness, and b^*: yellowness, ΔE: color difference. YCRO: Yellow corn rice oat; YCR: Yellow corn rice; WCRO: White corn rice oat; ROG: Rice oat gums. Different letters in the same column indicate significant differences ($p \leq 0.05$) between pasta for the same coordinate.

As it can be seen, the highest value of ΔE corresponds to WCRO pasta due to the low values of a^* and b^* parameters and the high value of L^* for the rest of the pasta. Further, the brightness (L^*) showed a significant increase ($p < 0.05$) in WCRO pasta concerning control pasta, while YCR and ROG did not present differences. On the other hand, the a^* value was significantly higher ($p < 0.05$) in YCRO pasta with a positive value, while the other pasta had negative values. Finally, the b^* parameter showed a greater significant difference ($p < 0.05$) in YCRO pasta, followed by YCR with respect to control and other gluten-free pasta.

3.3. Technological Properties

The technological quality parameters of gluten-free pasta enriched with seabass concentrate and control pasta are shown in Figure 1.

The weight gain (WG) is expressed on a scale from 0 to 140%; on the contrary, the moisture (M), the swelling index (SI), and the cooking losses (CL) are expressed on another scale that goes from 0 to 10%. The GW represented in percentage was similar between control pasta, YCRO, and YCR. The SI values showed that gluten-free pasta showed significant differences ($p < 0.05$) concerning control with variations from 1.56 to 1.72%. CL showed less significant differences ($p < 0.05$) in gluten-free pasta than in control pasta, except YCR, which had a larger significant difference. Furthermore, the moisture was around 8.6% for all pasta, with WCRO standing out with the highest value and YCRO with the lowest ($p < 0.05$) value compared to control pasta.

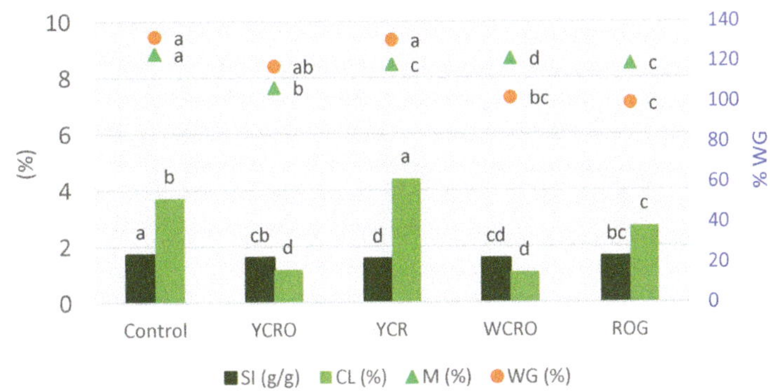

Figure 1. Technological properties of gluten-free and control pasta. Parameters: SI: swelling index, CL: cooking losses, M: moisture, WG: weight gain. Treatment: YCRO: Yellow corn rice oat; YCR: Yellow corn rice; WCRO: White corn rice oat; ROG: Rice oat gums. Different letters represent significant differences ($p < 0.05$) among pasta for each parameter analyzed.

3.4. Sensory Parameters

3.4.1. Sensory Texture Profiles (STP)

The sensory texture profile for gluten-free and control pasta evaluated by selected assessors, on a scale from 0 to 5, where 0 is not at all and 5 is a lot, are represented in the semantic differential graph shown in Figure 2.

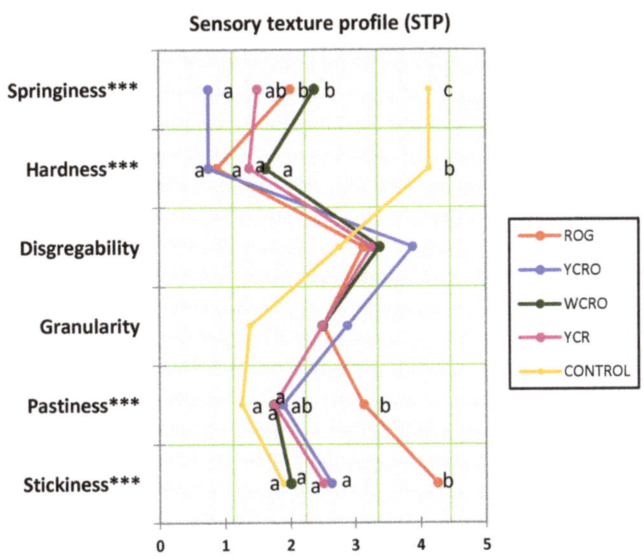

Figure 2. Sensory texture profile (STP) of developed pasta. YCRO: Yellow corn rice oat; YCR: Yellow corn rice; WCRO: White corn rice oat; ROG: Rice oat gums. Different letters represent significant differences ($p < 0.05$) between pasta for each attribute analyzed. *** Parameters with high significant differences ($p < 0.001$) among different pasta.

The parameters of springiness, hardness, pastiness, and stickiness resulted in significant differences ($p < 0.05$) between pasta. After analyzing the results of the studied texture attributes by the sensory panel and evaluating their performance, it was confirmed that they were capable of correctly discriminating the texture descriptors for each type of pasta.

All gluten-free pasta had similar behavior, except for the control durum pasta, which had a different composition concerning the other ones. The sensory analysis confirmed that the most important differences between gluten-free pasta and control were found in hardness and springiness according to TPA. In this way, the highest value of springiness was found in the control pasta, and the lowest value was found in YCRO. Regarding hardness, all gluten-free pasta had similar behavior, although they had significant differences with the control durum pasta. On the other hand, ROG had a higher pastiness and stickiness, showing significant differences with the other types of pasta.

3.4.2. Sensory Profile of Gluten-Free Pasta Developed (QDA)

A quantitative descriptive analysis, which is shown in Figure 3, was performed to characterize the pasta. The performance of the trained assessors' panel was checked by a panel analysis, not observing significant differences among judges' evaluations for all studied pasta. The plot from squares cosines study of measurements of trained assessors represented 97.52% of the total variation among kinds of pasta evaluated. The first component described the 81.32% and separated control pasta from gluten-free pasta. Control was characterized by its hardness, typical pasta aroma, typical pasta flavor, homogeneity, and the characteristic color, while gluten-free pasta was associated with fishy flavor and odor and a certain aftertaste. The second component with 16.20% discriminated in the upper part YCRO and YCR while the others (ROG and WCRO) were in the lower part, related to flavor and characteristic aroma of wheat pasta. YCRO was similar to YCR, while WCRO resembled ROG.

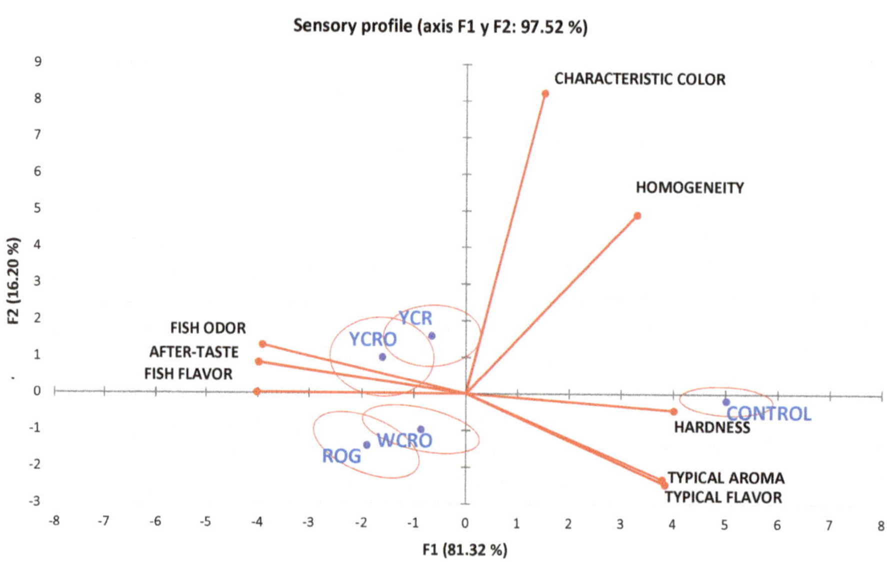

Figure 3. Sensory profile for enriched gluten-free pasta and control pasta. YCRO: Yellow corn rice oat; YCR: Yellow corn rice; WCRO: White corn rice oat; ROG: Rice oat gums.

3.5. *Comparative Study between Quality Parameters and TPA*

Table 4 shows the linear relationship between the parameters from TPA and technological properties in developed gluten-free pasta. A Pearson correlations analysis was carried out to check associations and significances.

Table 4. Coefficient correlations (r^2) between quality parameters and TPA.

TPA	WG (%)	SI (g/g) *	CL (%)	M (%)
HARDNESS	0.714	0.872	0.517	0.406
ADHESIVENESS	0.555	0.119	−0.174	−0.115
SPRINGNESS	0.696	0.840	0.471	0.470
COHESIVENESS	0.433	0.854	0.006	0.233
GUMMINESS	0.624	0.904	0.366	0.388
CHEWINES	0.621	0.901	0.391	0.431
FRACTURABILITY	0.819	0.777	0.662	0.397

WG: weight gain; SI: swelling index; CL: cooking losses; M: moisture. * Statistical significance ($p < 0.05$).

As can be seen, the only parameter of technological properties related significantly ($p < 0.05$) with the TPA parameters was the swelling index. Highlighted the relationship between this index and gumminess and chewiness parameters, all of them related to the viscoelastic behavior of materials.

3.6. Comparative Study between TPA and STP

Due to the great correlations in some cases ($r^2 > 0.881$), a principal component analysis (Figure 4) was performed to obtain a global version of the study and observe the relations between TPA and STP for all types of pasta.

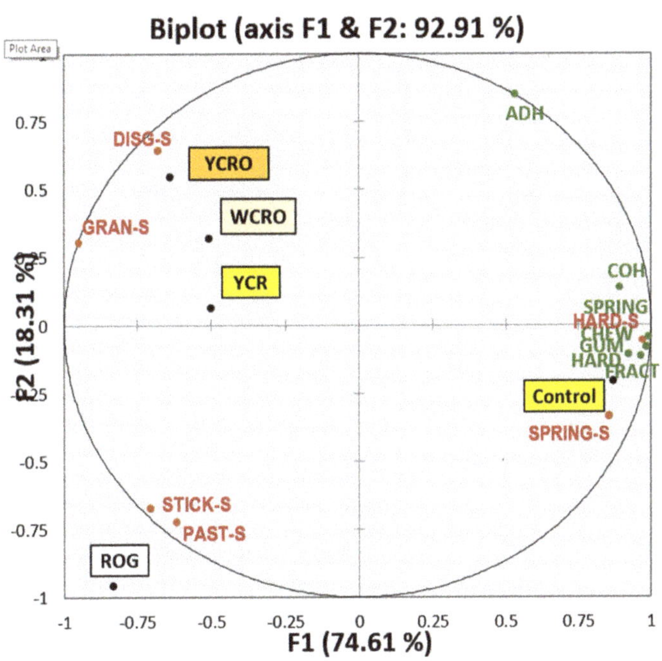

Figure 4. Principal component analysis of texture for enriched gluten-free pasta. HARD: hardness; AHD: adhesiveness; SPRING: springiness; COH: cohesiveness; GUM: gumminess; CHEW: chewiness; FRACT: fracturability; HARD.S: sensorial hardness; DISG.S: sensorial disintegration; GRAN.S: sensorial graininess; SPRING.S: sensorial springiness; PAST.S: sensorial pastiness; STICK.S: sensorial stickiness; YCRO: yellow corn + rice + oat; WCRO: white corn + rice + oat; YCR: yellow corn + rice; ROG: rice + oat + gum.

The PCA described 92.91% of the variability of the study. The first component was 74.61% and separated control pasta from gluten-free pasta. This allowed finding differences in texture between gluten-free pasta and control pasta. YCRO, YCR, and WCRO were characterized by sensorial disintegration and graininess. On the other hand, ROG was characterized by sensorial pastiness and stickiness, which had a negative correlation with instrumental adhesiveness. Control pasta was associated with the most instrumental parameters (springiness, cohesiveness, fracturability, gumminess, chewiness, hardness, and adhesiveness) and sensorial hardness. The second component, with 18.31%, allowed us to separate treatment by the main ingredient of pasta.

4. Discussion

4.1. Texture Study of Gluten-Free Pasta Enriched with Fish

As seen in the results, gluten-free pasta had lower cooking times than the control pasta—this could be a consequence of the incorporation of other ingredients. Due to durum wheat was not used in pasta-making, the starch content of the final product and the water required for its gelatinization could be decreased. Further, since it does not contain gluten, the main proteins that form the original structure of traditional pasta decrease when replaced by lower molecular weight and, therefore, require less time to hydrate [25].

Regarding texture profile analysis (TPA), the hardness of pasta is determined by the matrix formed by the gluten network during cooking, in cooperation with other substances such as lipids and starch [26]. The gluten-free pasta did not have this traditional viscoelastic network as it was made from other cereals different from wheat with little or no prolamine in its composition. In addition to the above, the fact that it had in its composition fish concentrate that introduced myofibrillar proteins and fat produced a change of the structure of the dough, causing a weakening in the final product. This modification is because the new ingredients modified the three-dimensional network in the gluten-free pasta, changing the nature and behavior of the material, and as a consequence, modifying the textural properties of the dough, causing mainly hardness [27]. For its part, the adhesiveness increased too due to the weakening of the structure by the change of cereals and the gum addition to improve the consistency. As said before, the new product does not have gluten-forming proteins in its composition. Thus, with the incorporation of fish, rich in myofibrillar proteins and lipids, the observed decrease in springiness due to the absence of wheat proteins is understandable [28].

The texture profiles of gluten-free pasta analyzed by instrumental methods showed important changes in their rheological parameters. Highlighted the addition of fish to pasta because it interfered with proper gel formation and, therefore, affected the interaction between starch and cereal proteins, modifying the texture in the final products and increasing the amount of free water [29]. This led to a major change in the rheological behavior of enriched gluten-free pasta developed in this experience.

4.2. Color of Developed Pasta

The yellow index (b^*) is one of the most important parameters in pasta acceptability due to the characteristic color of pasta [30]. YCRO and YCR pasta, both made with yellow corn, showed higher values of the b^* parameter, which could be due to the more intense yellow color provided by this cereal in comparison to durum wheat. All enriched pasta developed in this work showed an appreciable variation with respect to the control, as shown by ΔE, due to the characteristic color of cereals used in each pasta-making formulation. Our findings are in agreement with a previous study which showed that pasta with sorghum, suitable for people with celiac disease, had values of ΔE with an increase from 20% to 40% [31].

4.3. Technological Properties of Developed Pasta

Regarding weight gain, the highest values were found in control wheat pasta and in the yellow corn pasta, which presents a higher amount of amylopectin than the rest of the cereals used. For this reason, these pastas tend to hold more water and swell more [25]. On the other hand, weight gain is reduced by the addition of fish because fish proteins and lipids interact and compete with starch for the absorption of water during cooking, reducing its hydration and gelation [12]. Concerning the swelling index, gluten-free pasta absorbed less water and, therefore, had a decrease in the size of the starch granule responsible for its gelatinization [25,32].

Regarding cooking losses, YCR showed a great difference from the control. This pasta was made without oat bran which differentiates it from the rest. Thus, it could be deduced that oat bran is an ingredient that causes a decrease in cooking losses. Regarding the latter, there is some controversy about the effect of fiber on this technological property. Some studies affirm that enriching pasta with fiber reduces losses since they contribute to the development of the protein matrix by fortifying the structure of pasta [33,34], while others indicate that fiber has greater water absorption capacity and interferes in the formation of the structure and cooking losses will increase [35]. In the same way, the key technological parameter moisture showed that all pasta had values below those established by the BEDCA (9.5%) and Spanish regulation [1,36]. Thus, it could be said that pasta has been made following a correct procedure, in which the applied drying allowed to standardize the moisture content achieve extending pasta gluten-free shelf life.

4.4. Sensory Study (Characterization of Product)

According to the sensory texture profile (STP), the principal differences among gluten-free pasta and control pasta were in hardness and springiness; this result coincided withs those provided by the instrumental method (TPA and WB). Pasta made according to ROG formulation was found to be significantly the stickiest and pastiest one. This may be because, in their composition, only rice and oats were used as substitutes for durum wheat. These ingredients modify the rheological properties due to their content of starch and soluble fiber, especially in the case of oats.

As can be seen in the results of the sensory profiles (Figure 3) obtained from QDA, YCRO pasta was similar to YCR treatment, while WCRO was similar to ROG pasta. All gluten-free pasta with fish added made in this study was mainly characterized by a moderate fish odor, a certain aftertaste, and perceivable fish flavor above the typical wheat pasta smell. In the same order of ideas, color and smell in the products were consistent since they coincide with the characteristics that each of the ingredients used in the different formulations is supposed to attribute. At the same time, the control pasta is distinct, as was to be expected, by a typical aroma (cooked pasta) and flavor, with appropriate hardness with characteristic and homogeneous color.

To assess those attributes that must be in an enriched pasta with fish, a comparison was made with the results of a previous study in durum wheat pasta with a fish concentrate where in general, it was characterized by a typical yellow color, farinaceous smell, and typical semolina flavor [37]. Taking all of the above into consideration, the gluten-free pasta enriched with fish that most closely resembles these characteristics is the YCR formulation.

4.5. Technological Quality Parameters and TPA

A comparative study demonstrated that even though all quality parameters showed high correlation indices (r^2), only the swelling index (SI) was statistically significant ($p < 0.05$). This close relationship may have been due to the tendency to increase both aspects, i.e., texture parameters (hardness, gumminess, and chewiness) and technological index. As is known, SI of pasta is an indicator of water absorbed by the starch and proteins during cooking which is utilized for starch gelatinization and protein hydration. Several authors reported that the water absorption capacity depends on the behavior of the proteins denaturation and the function of the amylose/amylopectin ratio, as well as the chain length distribution of amylopectin [38,39]. The above, together with the amylaceous nature of the cereals used to replace wheat semolina, could be the reason that particularly parameters of texture such as hardness, gumminess, and chewiness could have resulted very sensible to amount of water retained in the structural net of the dough.

4.6. Global Approach of Texture Study in Gluten-Free Pasta Enriched with Fish

A comparative study between TPA and STP results was carried out, as shown in Figure 4. The principal component analysis takes into account the results for texture from the sensory evaluation and instrumental analysis, considering, at the same time, all kinds of pasta assayed (experimental gluten-free and control pasta) were represented in a unique biplot that collected 92.91% of the total variation of the study. The above fact provided reliability and robustness to the results of the research. In this sense, it is evident how the STP was capable of discriminating of an adequate form both the attributes and the types of pasta analyzed, in contrast with the values obtained for the different determined parameters in TPA, which were located around the control pasta, demonstrating a low discrimination capacity concerning the rest of the treatment with the exception of the adherence located away from the rest of the instrumental parameters of texture.

It could be claimed that although, in theory, a TPA attempts to mimic what happens in the mouth during the chewing of food [21], it is not as effective in discriminating between traits and attributes as an objective quantitative–descriptive sensory analysis. However, the instrumental determination, as we have already seen, correlates very well with technological parameters, which, together with the information obtained on the organoleptic properties of each type of pasta developed, allows us to have an overall view of the study to have a global understanding and make decisions more accurately.

5. Conclusions

The use of cereals other than wheat in enriched pasta with fish could represent a good alternative to contribute to achieving the health of the consumer, especially in the coeliac and/or gluten intolerant population. Gluten-free pasta enriched with fish requires less cooking time than wheat pasta. Further, an effect of oat bran could be seen over the texture properties. The use of different ingredients led to a weakening of the structure of gluten-free pasta and, therefore, to less hardness, springiness, gumminess, chewiness, and fracturability. Concerning color, the addition of yellow corn gave gluten-free pasta a similar color to durum wheat pasta (control). The TPA and STP coincided in pointing out springiness and hardness as the parameters that differentiate the gluten-free pasta from the control pasta. Regarding sensory analysis, all gluten-free pasta were characterized by fish aromas and flavors with a certain aftertaste and a lower hardness in comparison with durum pasta. Finally, according to technological, physical, and sensory parameters, gluten-free pasta made with yellow corn and rice was the most similar to the control pasta.

Author Contributions: Conceptualization, A.A., A.V. and J.B.C.M.; Data curation, A.A., A.V., J.A.B.G. and J.B.C.M.; Formal analysis, A.A., A.V., J.A.B.G. and J.B.C.M.; Funding acquisition, J.A.B.G. and J.B.C.M.; Investigation, A.A., A.V., P.M., J.A.B.G. and J.B.C.M.; Methodology, A.A., A.V., A.H., P.M. and J.B.C.M.; Project administration, A.A., P.M., J.A.B.G. and J.B.C.M.; Resources, J.A.B.G.; Software, A.A and J.B.C.M.; Supervision, J.A.B.G. and J.B.C.M.; Validation, J.B.C.M.; Visualization, J.B.C.M.; Writing—original draft, A.A., A.V. and A.H.; Writing—review and editing, A.A, P.R., J.A.B.G. and J.B.C.M. All authors have read and agreed to the published version of the manuscript.

Funding: This project was subsidized by project OTRI-UNIZAR within the framework of the I convocatoria Valle de la Muerte.

Data Availability Statement: The datasets generated for this study are available on request to the corresponding author.

Acknowledgments: The authors thank the companies PASTAS ROMERO and SCANFISK for the contribution in raw material and technical support.

Conflicts of Interest: The authors declare no conflict of interest.

References

1. *Real Decreto 2181/1975, de 12 de Septiembre, por el que se Aprueba la Reglamentación Técnico-Sanitaria para la Elaboración;* Circulación y Comercio de Pastas Alimenticias; Boletín Oficial del Estado No 220: Madrid, Spain, 1975.
2. Caio, G.; Volta, U.; Sapone, A.; Leffler, D.A.; De Giorgio, R.; Catassi, C.; Fasano, A. Celiac disease: A comprehensive current review. *BMC Med.* **2019**, *17*, 142. [CrossRef]
3. Husby, S.; Koletzko, S.; Korponay-Szabo, I.; Kurppa, K.; Mearin, M.L.; Ribes-Koninckx, C.; Shamir, R.; Troncone, R.; Auricchio, R.; Castillejo, G.; et al. European Society Paediatric Gastroenterology, Hepatology and Nutrition Guidelines for Diagnosing Coeliac Disease 2020. *J. Pediatr. Gastroenterol. Nutr.* **2020**, *70*, 17. [CrossRef] [PubMed]
4. ¿Qué es la Enfermedad Celiaca? Available online: https://celiacos.org/enfermedad-celiaca/que-es-la-enfermedad-celiaca/ (accessed on 4 November 2021).
5. Reglamento 828/2014 de la Comisión—De 30 de Julio de 2014—Relativo a los Requisitos Para la Transmisión de Información a los Consumidores Sobre la Ausencia o la Presencia Reducida de Gluten en los Alimentos. Available online: http://data.europa.eu/eli/reg_impl/2014/828/oj (accessed on 15 November 2021).
6. Marti, A.; Pagani, M.A. What can play the role of gluten in gluten free pasta? *Trends Food Sci. Technol.* **2013**, *31*, 63–71. [CrossRef]
7. Flores, M.V.; Vera, N.G.; Lira, A.Q.; Martini, J.P. La sustitución parcial o total de sémola de trigo en las pastas y su valor nutricional. *Boletín Cienc. Agropecu. ICAP* **2017**, *3*. [CrossRef]
8. Molina-Rosell, C. Alimentos sin gluten derivados de cereales. In *Enfermedad Celíaca y Sensibilidad al Gluten no Celíaca;* Rodrigo, L., Peña, A.S., Eds.; OmniaScience: Terrassa, Spain, 2013; pp. 447–461. ISBN 978-84-940234-3-9. [CrossRef]
9. Ainsa, A.; Marquina, P.L.; Roncalés, P.; Beltrán, J.A.; Calanche, J.B. Enriched Fresh Pasta with a Sea Bass By-Product, a Novel Food: Fatty Acid Stability and Sensory Properties throughout Shelf Life. *Foods* **2021**, *10*, 255. [CrossRef]
10. Anbudhasan, P.; Asvini, G.; Surendraraj, A.; Ramasamy, D.; Sivakumar, T. Development of Functional Pasta Enriched with Omega-3 Fatty Acids. *Fish Technol.* **2014**, *51*, 242–246.
11. Calanche, J.; Beltrán, H.; Marquina, P.; Roncalés, P.; Beltrán, J.A. Eating fish in another way: Development of functional pasta with added concentrates of farmed sea bass (*Dicentrarchus labrax*). *Cereal Chem.* **2019**, *96*, 856–865. [CrossRef]
12. Desai, A.S.; Brennan, M.A.; Brennan, C.S. Influence of semolina replacement with salmon (*Oncorhynchus tschawytscha*) powder on the physicochemical attributes of fresh pasta. *Int. J. Food Sci. Technol.* **2018**, *54*, 1497–1505. [CrossRef]
13. Monteiro, M.L.G.; Mársico, E.T.; Deliza, R.; Castro, V.S.; Mutz, Y.S.; Soares Junior, M.S.; Caliari, M.; dos Santos, E.A.; Conte-Junior, C.A. Physicochemical and sensory characteristics of pasta enriched with fish (Oreochromis niloticus) waste flour. *LWT* **2019**, *111*, 751–758. [CrossRef]
14. Berruezo, G.; Martínez, C. Calidad y composición nutritiva de la carne, el pescado y el marisco. *Tratado Nutr.* **2005**, *2*, 107–146.
15. Ainsa, A.; Iranzo, L.; Honrado, A.; Marquina, P.; Roncalés, P.; Beltrán, J.A.; Calanche, J. Effects of cooking over the stability of fatty acids as bioactive compounds in enriched pasta with a fish by-product. *Cereal Chem.* **2021**. [CrossRef]
16. Martínez, O. Estado actual del aprovechamiento de subproductos de la industria pesquera mediante la obtención de productos de alto valor añadido. *Rev. Tecnol. Hig. Aliment.* **2011**, *429*, 71–80.
17. Ficco, D.B.M.; De Simone, V.; De Leonardis, A.M.; Giovanniello, V.; Del Nobile, M.A.; Padalino, L.; Lecce, L.; Borrelli, G.M.; De Vita, P. Use of purple durum wheat to produce naturally functional fresh and dry pasta. *Food Chem.* **2016**, *205*, 187–195. [CrossRef] [PubMed]
18. AACC. *AACC Approved Methods of the American Association of Cereal Chemists*, 10th ed.; AACC: St. Paul, MN, USA, 2000.
19. Cleary, L.; Brennan, C. The influence of a (1→3)(1→4)-β-d-glucan rich fraction from barley on the physico-chemical properties and in vitro reducing sugars release of durum wheat pasta. *Int. J. Food Sci. Technol.* **2006**, *41*, 910–918. [CrossRef]
20. ISO 11036:2020. *Sensory Analysis. Methodology. Texture Profile;* International Organization for Standarization: Geneva, Switzerland, 2020.
21. Szczesniak, A.S. Texture is a sensory property. *Food Qual. Prefer.* **2002**, *13*, 215–225. [CrossRef]
22. UNE-EN ISO. *Análisis Sensorial. Guía General para la Selección, Entrenamiento y Control de Catadores y Catadores Expertos. ISO 8586:2012;* International Organization for Standarization: Genova, Switzerland, 2014.
23. UNE-EN ISO. *Análisis Sensorial. Metodología. Guía General. ISO 6658:2017;* International Organization for Standarization: Genova, Switzerland, 2019.
24. Di Pede, G.; Dodi, R.; Scarpa, C.; Brighenti, F.; Dall'Asta, M.; Scazzina, F. Glycemic Index Values of Pasta Products: An Overview. *Foods* **2021**, *10*, 2541. [CrossRef]
25. Dergal, S.B.; Rodriguez, H.B.; Morales, A.A. Química de los Alimentos, 4th ed.; 2006. Available online: http://madejaseinsumos.com/descargables/Quimica%20de%20los%20alimentos.pdf (accessed on 15 November 2021).
26. Larrosa, V.; Lorenzo, G.; Zaritzky, N.; Califano, A. Improvement of the texture and quality of cooked gluten-free pasta. *LWT* **2016**, *70*, 96–103. [CrossRef]
27. Lara, F.V.; Amat, S.V.; Islas, A.R.; Baviera, J.M.B.; Meló, R.G.; Casillas, R. Efecto de la sustitución de harina de trigo con harina de avena, maíz y sorgo sobre las propiedades reológicas de la masa, texturales y sensoriales del pan E ect of substitution of wheat our with oat, corn and sorghum on dough rheological, textural and sensorial bread properties. *Rev. Iberoam. Tecnol. Postcosecha* **2016**, *17*, 307–317.
28. Atwell, W.A. *Wheat Flour*, 2nd ed.; American Association of Cereal Chemists: St. Paul, MN, USA, 2016.

29. Yang, H.; Zhang, W.; Li, T.; Zheng, H.; Khan, M.A.; Xu, X.; Sun, J.; Zhou, G. Effect of protein structure on water and fat distribution during meat gelling. *Food Chem.* **2016**, *204*, 239–245. [CrossRef]
30. Pongpichaiudom, A.; Songsermpong, S. Evaluation of microstructure and quality characteristics of microwave-dried instant noodles enriched with chicken meat, egg yolk, and seaweed. *J. Food Meas. Charact.* **2018**, *12*, 22–34. [CrossRef]
31. Khan, I.; Yousif, A.M.; Johnson, S.K.; Gamlath, S. Effect of Sorghum Flour Addition on in vitro Starch Digestibility, Cooking Quality, and Consumer Acceptability of Durum Wheat Pasta: Effect of sorghum flour addition. *J. Food Sci.* **2014**, *79*, S1560–S1567. [CrossRef] [PubMed]
32. Palavecino, P.M.; Bustos, M.C.; Heinzmann Alabí, M.B.; Nicolazzi, M.S.; Penci, M.C.; Ribotta, P.D. Effect of Ingredients on the Quality of Gluten-Free Sorghum Pasta: Gluten-free sorghum flour pasta. *J. Food Sci.* **2017**, *82*, 2085–2093. [CrossRef]
33. Mert, B.; Tekin, A.; Demirkesen, I.; Kocak, G. Production of Microfluidized Wheat Bran Fibers and Evaluation as an Ingredient in Reduced Flour Bakery Product. *Food Bioprocess Technol.* **2014**, *7*, 2889–2901. [CrossRef]
34. Sabanis, D.; Tzia, C. Effect of hydrocolloids on selected properties of gluten-free dough and bread. *Food Sci. Technol. Int.* **2011**, *17*, 279–291. [CrossRef] [PubMed]
35. Sivam, A.S.; Sun-Waterhouse, D.; Quek, S.; Perera, C.O. Properties of Bread Dough with Added Fiber Polysaccharides and Phenolic Antioxidants: A Review. *J. Food Sci.* **2010**, *75*, R163–R174. [CrossRef] [PubMed]
36. BEDCA. Base de Datos Española de Composición de Alimentos. 2019. Available online: https://www.bedca.net/bdpub/ (accessed on 20 October 2021).
37. Ainsa, A.; Honrado, A.; Marquina, P.L.; Roncalés, P.; Beltrán, J.A.; Calanche, M.J.B. Innovative Development of Pasta with the Addition of Fish By-Products from Two Species. *Foods* **2021**, *10*, 1889. [CrossRef]
38. Granito, M.; Guerra, M.; Torres, A.; Guinand, J. Efecto del procesamiento sobre las propiedades funcionales de Vigna Sinensis. *Interciencia* **2004**, *29*, 521–526.
39. Köber, E.; Gonzalez, M.E.; Gavioli, N.; Salmoral, E.M. Modification of water absorption capacity of a plastic based on bean protein using gamma irradiated starches as additives. *Radiat. Phys. Chem.* **2007**, *76*, 55–60. [CrossRef]

foods

MDPI

Article

Date, Apple, and Pear By-Products as Functional Ingredients in Pasta: Cooking Quality Attributes and Physicochemical, Rheological, and Sensorial Properties

Brahim Bchir [1,2,*], Romdhane Karoui [3], Sabine Danthine [4], Christophe Blecker [4], Souhail Besbes [1] and Hamadi Attia [1]

1 Laboratory of Analysis Valorization and Food Safety, National Engineering School of Sfax, University of Sfax, Sfax BP W-3038, Tunisia; besbes_souhail@yahoo.fr (S.B.); hamadi.attia@gmail.com (H.A.)

2 Higher Institute of Biotechnology of Monastir, University of Monastir, Avenue Taher Hadded BP 74, Monastir 5000, Tunisia

3 Univ. Artois, Univ. Lille, Univ. Littoral Côte D'Opale, Univ. Picardie Jules Verne, Univ. de Liège, INRAE, Junia, UMR-T 1158, BioEcoAgro, F-62300 Lens, France; romdhane.karoui@univ-artois.fr

4 Laboratory of Food Science and Formulation, Gembloux Agro-Bio Tech, University of Liège, Passage des Déportés 2 B, B-5030 Gembloux, Belgium; sabine.danthine@uliege.be (S.D.); christophe.blecker@uliege.be (C.B.)

* Correspondence: bchbrahim@yahoo.fr; Tel.: +216-53-440-380

Citation: Bchir, B.; Karoui, R.; Danthine, S.; Blecker, C.; Besbes, S.; Attia, H. Date, Apple, and Pear By-Products as Functional Ingredients in Pasta: Cooking Quality Attributes and Physicochemical, Rheological, and Sensorial Properties. *Foods* **2022**, *11*, 1393. https://doi.org/10.3390/foods11101393

Academic Editors: Laura Gazza and Francesca Nocente

Received: 2 October 2021
Accepted: 31 October 2021
Published: 12 May 2022

Abstract: This study aims to evaluate the impact of incorporating pear, date, and apple by-products on pasta properties. Pasta properties including cooking quality, texture, color, rheology, thermal gelling, and microstructural characteristics were evaluated. Common wheat flour was substituted by 0, 2.5, 5, 7, and 10 g/100 g of by-products. To choose the best-suited substitute of flour for the preparation of pasta, the sensorial properties of pasta were investigated. Interrelationships between all the physicochemical parameters were investigated using multiple factor analysis. We also studied the impact of storage (7, 15, and 30 days) on the physicochemical proprieties of pasta. The results revealed that the chemical composition of pasta elaborated with by-products was characterized by higher energy (~386 Kcal) and fiber content (~13%) than the control pasta. Generally, materials added to the durum wheat pasta reduce optimum cooking time, adhesiveness, and extensibility, and enhance the swelling index, cooking loss, cooking water absorption, water activity, firmness, and tenacity of pasta. Cooked pasta samples were significantly ($p < 0.05$) darker (L*) and greener (-a*) than the control pasta. Increasing the rate of by-products from 2.5% to 10% principally altered the texture and structure of pasta. Scanning electron microscopy analysis showed that the inclusion of by-products into pasta leads to a disruption of the protein matrix. A practical formulation (2.5% of by-products) can be selected, since a significant difference was detected between overall acceptability scores. Grouping the variables in the principal component analysis plot showed that pasta samples can be divided into three groups. Each group was correlated by a specific variable. A significant modification of the physical parameters of pasta was observed after 30 days of storage.

Keywords: pear, date, and apple by-products; pasta; wheat flour; cooking quality; sensorial properties; scanning electron microscopy; multiple factor analysis

1. Introduction

Health authorities worldwide recommend reducing the consumption of animal proteins and fats and increasing cereal intake, which is a crucial source of dietary fibers [1]. In addition, the World Health Organization considers pasta (a cereal product) as a suitable vehicle for the supplementation of nutrients [2]. Pasta is a traditional food produced using durum wheat flour, and is a staple food in many countries. It is favored by consumers for its ease of transportation, cooking, handling, and storage, as well as its low cost and low glycemic index (2–4) [3]. Nevertheless, traditional pasta is claimed to lack other essential nutritional components such as dietary fibers, vitamins, proteins, minerals, and many other valuable healthy components, such as bioactive molecules [3]. In reality, consumers demand foods with traditional nutritional aspects and additional health benefits.

For the preparation of pasta, it is possible to use non-durum wheat flour and other ingredients (such as dietary fibers) to produce special pasta. By combining the benefits of pasta with the benefits of dietary fibers, new functional food products are created to prevent and to treat diseases such as diabetes and coronary heart disease. Thus, a formulation that supplements pasta with a higher dietary fiber content could enhance fiber intake and decrease the glycemic index of pasta [3]. Different materials have been used as a source of dietary fibers in the preparation of pasta. These materials contain legume flours (bean, chickpea, soybean, pea, and lentil) and flour from other cereals such as amaranth maize, barley, oat, rice, and sorghum [2–6]. The incorporation of dietary fibers into pasta affects the integrity of the protein starch network, hence the tenacity of protein–starch products, and increases pasta quality in terms of optimum cooking time, water absorption, swelling index, texture, taste, appearance, and cooking loss [7].

Some studies suggest adding dietary fruit fiber by-products to wheat flour in pasta formulations due to their nutritional and functional properties, as well as reduced risks of industrial environmental contamination, lower cost, and health benefits [8]. Indeed, different by-products have been used in the formulation of wheat spaghetti, such as tomato and potato pulp by-products [9], carob fibers, carrot pomace [10], unripe banana, and plantain flour [11]. Aguedo et al. [12] showed that the by-products of cooked fruit have specific aromas and high fiber content.

In Belgium, the fabrication of 'Liege syrup', a popular fruit concentrate, generates 1000 t of by-products annually from cooked dried dates, pears, and apples [12–14]. Agri-food industry by-products are an environmental problem and have repercussions on society and the economy [1]. By-product valorization allows novel ingredients with great nutritional value and healthy properties to be obtained, which can be a potent strategy with which to respond to the demand of consumers for healthier processed foods and to decrease waste [1]. Thus, the development of new products, such as pasta with by-products, can be a strategic area of the food industry. To the best of our knowledge, date, pear, and apple by-products have not been used for the formulation of pasta.

Therefore, the main objectives of this study are to (a) evaluate the impact of incorporating pear, apple, and date by-products on pasta properties, (b) select the best-suited substitute flour for the preparation of pasta, and (c) determine the impact of the storage period test (7, 15, and 30 days) on the physicochemical proprieties of pasta. Multiple factor analysis was used as a statistical tool to analyze the interrelationships between the physicochemical parameters of pasta.

2. Materials and Methods

2.1. Materials

The basic ingredients used for the formulation of pasta were commercial semolina flour (moisture: 10%; carbohydrate: 78%; protein: 13%; and lipid: 2%), water, and apple, pear, or date by-products. By-products came from dried fruit pomaces, either pears (French and Belgian 'Conference' cultivars), apples (Belgian 'Jonagold' and 'Jonagored', French 'Granny Smith' cultivars), or dates ('Deglet Noor' cultivar), from the fabrication of 'Liege syrup' from the Siroperie Meurens (Belgium). The pomaces were oven-dried for 7 h at 70 °C in a laboratory-scale dryer (Schutzart, Germany, Memmert tcp 800), and ground in a Fritsch laboratory mill with a 1 mm mesh sieve (Haan, Germany). The conservation method and the industrial process of by-products were shown in our previous work [15].

The physicochemical properties of date, apple, and pear by-products are presented in our preceding investigation [15]. The results revealed a predominance of fiber (82% < % of fiber < 91%) in all pomaces, especially insoluble fibers (78% < % of insoluble fibers < 89%), followed by protein (6% < % of protein < 11%), fat (2.5% < % of fat < 3.7%), ash (0.9% < % of ash < 1.4%), and free sugar (0.3% < % of free sugar < 1.1%). Regarding fiber content, it is noteworthy that by-products can be referred to as sources of fiber.

2.2. Pasta Manufacturing

The pasta was manufactured according to the method presented by Bouacida et al. [16]. Durum wheat pasta was made using commercial hard wheat semolina (La rose Blanche, Sousse, Tunisia), water, and different types of by-products (apple, pear, or date by-products). By-products were incorporated into recipes by replacing durum wheat flour at the following proportions (*w*/*w*): 2.5%, 5%, 7%, and 10% (*w*/*w*). An additional sample with no by-products included was also made as a control. All dried components of the formula (flour by-products: 70%) were mixed; water (30%) was then added in a Kenwood mixer (Serial, KM 336, Germany) at a short speed for 10 min until the 'dough' had an adequate consistency for lamination. After a rest of 1 h, the 'dough' was mixed for 10 min at a short speed. The dough was laminated (until it was 2 mm thick) and cut into strips approximately 5 mm wide and 15 cm long using a home-scale-sized pasta lamination machine (LUSSO SP 150, Italy). Then, pasta was dried using a dryer (Memmert tcp 800, Schutzart, Germany) at 45 °C to reach 13 m/m% moisture content, which is recommended by the Codex Alimentarius [17]. The samples were wrapped in cling film and conserved in airtight containers at room temperature for 30 days. The measurements were realized at selected time intervals (at 0, 7, 21, and 30 days).

2.3. Chemical Analysis and Nutritional Values

The chemical composition of the pasta was determined by AOAC (1997) methods for ash, lipid, and moisture. The amount of protein was determined using a Dumas Elementar Rapid N cube 161 15054 (Donaustrasse, Germany), as shown in our previous work [14]. The amount of dietary fiber was determined by a theoretical calculation, considering the quantity added to the sample. The amount of carbohydrate was estimated by the difference in mean values, 100-(sum of percentages of ash, moisture, protein, and lipid) [18]. Energy values were evaluated by using the factors 4, 4, and 9 for each gram of protein, ash, and lipid, respectively.

2.4. Physical Analysis

2.4.1. Cooking Properties

To determine the cooking properties, pasta samples were analyzed for their optimal cooking time, swelling index, cooking water absorption, and cooking loss. All tests were realized in triplicate.

Optimum Cooking Time

The optimum cooking time (OCT) is the time require to reach a complete gelatinization of starch [19]. According to method 16–50 (AACC, 2000), the OCT was determined as the time when the white inner core of the pasta disappeared after cross-cutting it with a razor blade, or after compressing the pasta between two glass slides during 30 s intervals [16].

Swelling Index

The swelling index (SI) of cooked pasta (grams of water per gram of dry pasta) was determined by drying pasta samples to a constant weight at 105 °C, expressed as Equation (1):

$$SI = \frac{W_1 - W_3}{W_3} \quad (1)$$

where W_1 is the weight of cooked product and W_3 is the weight after drying [20].

Cooking Water Absorption

Cooking water absorption (CWA) corresponds to the amount of water that a known dry pasta weight absorbs during cooking and holds after draining (Equation (2)):

$$\text{CWA}\ (\%) = \frac{W_1 - W_2}{W_2} * 100 \tag{2}$$

where W_1 is the weight of cooked product and W_2 is the weight of raw pasta [16].

Cooking Loss

Cooking loss (CL) is the quantity of dry matter lost in the cooking water under optimal cooking conditions [21]. CL was evaluated by evaporation to a constant weight in an air oven at 105 °C. The residue was weighed and reported as the percentage of the original pasta sample [4].

2.4.2. Quality Measurements

The effect of storage time on pasta qualities was determined during conservation in cling film and stored in airtight containers at room temperature (25 °C). The experiment lasted 4 weeks; samples were taken after 7, 21, and 30 days. During these periods texture, color, and water activity were measured.

Texture Measurements

The measurement of the texture of cooked pasta was carried out using a Texture Analyser TAXT2i (Stable Micro System, Watford, UK). The cooked pasta was submitted to compression testing, as shown by Borneo and Aguirre [5]. All pastas were cooked on the day of evaluation. Before testing the pastas, overflow of water was blotted with absorbent paper. Firmness and adhesiveness were calculated by performing a compression test using an AP/36 cylinder probe. The probe compresses the pasta sample by 75% of its original height. The operating conditions of the instrument were as follows: 2 mm/s post-test speed, 2 mm/s test speed, 2 mm/s pre-test speed, and 0.10 N trigger force. Adhesiveness is the force necessary to overcome the attractive force between the surface of the material with which the product comes into contact and the surface of the product, and firmness was determined as the maximum shear strength necessary for the rupture of a sample.

Color Measurements

Surface color measurements of pasta were measured according to the method shown by Bchir et al. [22], using a colorimeter (ColorFlex EZ, HunterLab, Reston, VA, USA). The pasta sample color is measured as chromatic ordinates L* (lightness), a* (redness–greenness), and b* (yellowness–blueness) values. Additionally, the total color difference (ΔE) between enriched pasta and control pasta was determined from L*, a*, and b* values.

Water Activity

Water activity (a_w) was measured using an Aqualab Cx-2 instrument (Decagon, Pullman, WA, USA) at 20 °C [22].

2.4.3. Rheological Characteristics

The impact of added by-products on dough rheology characteristics was carried out using an alveograph (Chopin AL 87, France). The 54-30-02 method [23] was employed to measure the alveograph test. The monitored parameters were the dough extensibility (L), the deformation energy (W), the tenacity or resistance to extension (P), and the curve configuration ratio (P/L ratio) of the dough [14].

2.4.4. Differential Scanning Calorimetry

Differential scanning calorimetry (DSC) was performed to determine the thermal gelling properties (gelatinization temperature; temperature onset of gelatinization: T_{onset};

gelatinization end point: T_{endset}; and total product enthalpy: ΔH) of raw pasta. Thus, to evaluate the influence that by-products might have on the properties of the starch fraction, a TA Instruments Q1000 DSC (New Castle, DE, USA) with a refrigerated cooling accessory and modulated capability was used. Indium and eicosane were used to calibrate the instrument (eicosane, T_{onset}: 36.8 °C, ΔH: 247.4 J g^{-1}; indium, T_{onset}: 156.6 °C, ΔH: 28.7 J g^{-1}). Specific heat capacity (Cp) was calibrated using a sapphire. The milled sample (0.5 mm) was first blended with distilled water (1:4) to a total weight of 15 ± 0.3 mg, then sealed hermetically in aluminum pans, and finally left to equilibrate for 1 h prior to the tests. An empty aluminum pan was used as a blank. The temperature range of the scan was 4 and 110 °C with a 10 °C/min heating rate [4].

2.5. Sensory Evaluation

The sensory analysis was performed by a hedonic test with 65 untrained participants (32 males and 33 females, aged 20–50 years), as described by Bchir et al. [15]. Each pasta was freshly cooked as per the procedure shown in Section 2.2. After cooking, the pastas were strained, rinsed, and cooled in water at 20 °C. Pastas (100 g) were served on odorless white paper plates with three-digit random number codes. A demographic survey was completed by the participants and evaluations were conducted using a seven-point hedonic scale to determine the degree to which the pasta was appreciated (7 = extremely appreciated, 4 = neither appreciated nor unappreciated, and 1 = extremely unappreciated). Samples were evaluated for the degree of to which they were appreciated for their appearance, color, taste, aftertaste, texture, and overall acceptability.

2.6. Scanning Electron Microscopy (SEM)

The microstructure of transversely fractured dough of pasta was performed by scanning electron microscopy (Thermoscientific, Q250, Cambridge, UK). The micrographs were taken using 200× magnification, 70 Pa pressure, and 15.00 KV high voltage.

2.7. Statistical Analysis

Statistical analyses were determined using a statistical software program (XLSTAT). Analysis of variance (ANOVA) was performed using Duncan's test to evaluate significant differences between the samples ($p < 0.05$). To classify the experimental samples of enriched pasta, multiple factor analysis was run using XLSTAT software 2018. Multiple factor analysis transforms the original measured variables into new, uncorrelated variables called principal components.

3. Results and Discussion

3.1. Chemical Composition of Pasta

The chemical compositions of pasta made with date, apple, and pear by-products are shown in Table 1. Pasta supplemented with date, pear, and apple by-products have a similar composition. Indeed, statistical analysis did not show a significant ($p > 0.05$) difference between pasta enriched with a similar rate of by-products. The chemical composition of pasta was characterized by a high percentage of carbohydrate followed by protein, ash, fiber, and fat. Adding by-product powders decreased the carbohydrate content (from 82 g to ~77%) and increased the fiber (from 4% to ~13%), protein (from 11.55% to ~12.12%), fat (from 1.65% to ~1.91%), and ash (from 4.31% to ~7.95%) fractions compared to the control.

The values of carbohydrate content in enriched pasta were higher than those obtained by Barbara et al. [2] for wheat spaghetti supplemented with silkworm flour (59.5 g/100 g). Protein values of tested pasta were slightly lower than those cited by Chillo et al. [24] and Bouacida et al. [16] for hard wheat spaghetti (13.5 g/ 100 g) and pasta enriched with Eruca leaves (18.74 g/100 g), respectively.

Table 1. Chemical composition of pasta and calculated nutritional values.

By-Product Addition (g/100 g Pasta)		Fiber (g/100 g)	Protein (g/100 g)	Carbohydrate (g/100 g)	Fat (g/100 g)	Ash (g/100 g)	Energy (Kcal)
Apple	2.5%	5.30 ± 0.01 [e]	11.79 ± 0.31 [a]	80.23 ± 1.10 [abc]	1.70 ± 0.21 [a]	6.28 ± 0.15 [de]	383.38 ± 3.22 [bc]
	5%	7.85 ± 0.25 [d]	11.99 ± 0.22 [a]	79.78 ± 1.15 [abc]	1.82 ± 0.21 [a]	6.41 ± 0.10 [d]	383.46 ± 1.15 [bcd]
	7.5%	10.08 ± 0.10 [c]	12.03 ± 0.51 [a]	78.59 ± 2.02 [bc]	1.86 ± 0.15 [a]	7.52 ± 0.50 [abc]	379.22 ± 2.54 [cde]
	10%	12.42 ± 0.20 [b]	12.12 ± 0.61 [a]	77.81 ± 2.50 [c]	1.88 ± 0.10 [a]	8.19 ± 0.20 [a]	376.64 ± 2.31 [e]
Pear	2.5%	5.60 ± 0.35 [e]	11.66 ± 0.55 [a]	80.36 ± 0.23 [abc]	1.71 ± 0.14 [a]	6.27 ± 0.22 [de]	383.47 ± 2.10 [bc]
	5%	7.29 ± 0.50 [d]	11.83 ± 0.10 [a]	79.18 ± 1.50 [bc]	1.80 ± 0.15 [a]	7.19 ± 0.25 [c]	380.24 ± 1.86 [cde]
	7.5%	10.28 ± 0.01 [c]	12.08 ± 1.02 [a]	78.59 ± 1.20 [bc]	1.91 ± 0.30 [a]	7.42 ± 0.70 [bc]	379.87 ± 3.18 [cde]
	10%	13.00 ± 0.61 [a]	12.01 ± 0.15 [a]	78.20 ± 1.15 [bc]	1.93 ± 0.01 [a]	7.77 ± 0.10 [abc]	378.57 ± 3.71 [de]
Date	2.5%	5.80 ± 0.55 [e]	11.79 ± 0.10 [a]	80.86 ± 2.15 [ab]	1.65 ± 0.21 [a]	5.70 ± 0.51 [e]	385.45 ± 2.50 [d]
	5%	7.45 ± 0.41 [d]	11.98 ± 0.21 [a]	79.80 ± 0.25 [abc]	1.70 ± 0.11 [a]	6.52 ± 0.45 [d]	382.42 ± 1.15 [bcd]
	7.5%	10.20 ± 0.10 [c]	12.05 ± 0.31 [a]	78.86 ± 2.50 [bc]	1.75 ± 0.15 [a]	7.34 ± 0.42 [abc]	379.39 ± 2.40 [cde]
	10%	12.70 ± 0.30 [ab]	12.11 ± 0.45 [a]	78.12 ± 1.15 [bc]	1.82 ± 0.40 [a]	7.95 ± 0.10 [ab]	377.30 ± 0.56 [de]
Control	0%	4.00 ±0.01 [f]	11.55 ± 0.60 [a]	82.49 ± 1.50 [a]	1.65 ± 0.20 [a]	4.31 ± 0.25 [f]	391.01 ± 1.23 [a]
F		**266.59**	**0.46**	**2.23**	**0.71**	**30.27**	**7.80**

Means in the same column with different letters are significantly different ($p < 0.05$).

According to Bouacida et al. [16], the value found for fiber (5.50 < % fiber < 13.00%) allows pasta to be classified as a product with a high content of fiber. This could be due to different ingredients, essentially from by-products, used in the formulation of pasta. The amount of fiber in supplemented pasta was in the range of that reported by Borneo and Aguirre [5], Aravind et al. [25], Bouacida et al. [16], and Minarovičová et al. [3], for wheat spaghetti enriched with spinach (4.12%), amaranth (5.79%), spinach leaves flour (4.16%), Eruca vesicaria leaves (10.9%), and pumpkin powder (27.2%), respectively. Table 1 shows that increasing the percentage of by-products in pasta significantly enhanced ($p < 0.05$) the amount of fiber, contrary to fat and protein values, which remained constant.

In addition, Table 1 reveals that fat contents were lower than those found in pasta presented by Borneo and Aguirre [5] (4.86 g/100 g) as well as Bouacida et al. [16] (4.16 g/100 g). This may be explained by the composition of by-products, which are characterized by a low amount of fat (2.5% < fat < 3.7%) [15], which can be an advantage for the food industry to produce low-fat foods. In fact, there is a tendency towards reducing the content of food constituents, such as cholesterol, salt, and fat, which have been related to human health concerns [26]. Table 1 shows that incorporating by-products significantly increased the rate of ash from 4.31% to ~7.95% in pasta ($p < 0.05$). This value is higher than those found by Shreenithee and Prabhasankar [27] in addition to Minarovičová et al. [3] for wheat spaghetti enriched with pea flour (2.56%), wheat flour (0.50%), and pumpkin powder (2.30%).

The incorporation of by-products into pasta reduces the energy of pasta compared to the control (Table 1). Therefore, pasta enriched with by-products could be introduced in the diet food plan. The energy provided by pastas in this work was in the range of that found by Barbara et al. [2] (366–369 Kcal).

3.2. *Physical Parameters of Pasta*

3.2.1. Cooking Properties of Pasta

Cooking properties are a major parameter for the judgment of pasta. Table 2 reveals a significant ($p < 0.05$) decrease in the optimum cooking time (OCT) (from 17.30 to 10.18 min) for all supplemented pasta samples compared to the control. However, the swelling index (SI), cooking water absorption (CWA), and cooking loss (CL) increased significantly ($p < 0.05$), from 2.54 to 3.86 g/g, 98.50 to 123.20, and 3.80 to 6.71 g/100 g, respectively, when the concentration of fiber increased (Table 2). This agrees with the fact that fiber by-products have a higher holding capacity than wheat flour used for the formulation of pasta [14,15]. Results for apple, pear, and date by-product-enriched pasta showed no significant difference ($p < 0.05$) in all cooking parameters when using the same rate of substitution.

Table 2. Cooking properties of enriched pasta.

By-Product Addition (g/100 g DM)		Optimum Cooking Time (Min)	Swelling Index (g of Water/g of Pasta)	Cooking Water Absorption (g/kg)	Cooking Loss (g/100 g of Pasta)
Apple	2.5%	16.20 ± 0.01 b	2.94 ± 0.12 de	111.00 ± 2.50 h	4.06 ± 0.15 cd
	5%	13.32 ± 0.04 f	3.05 ± 0.01 cd	114.10 ± 2.15 fg	4.55 ± 0.52 c
	7.5%	12.47 ± 0.04 g	3.25 ± 0.05 bc	120.32 ± 0.05 bcd	5.10 ± 0.11 b
	10%	10.51 ± 0.01 j	3.86 ± 0.04 a	123.20 ± 1.15 a	6.41 ± 0.16 a
Pear	2.5%	15.06 ± 0.02 d	2.82 ± 0.13 e	113.50 ± 1.50 gh	4.10 ± 0.10 cd
	5%	13.54 ± 0.03 e	2.95 ± 0.02 de	116.41 ± 2.12 ef	4.21 ± 0.13 c
	7.5%	12.26 ± 0.01 h	3.14 ± 0.11 bcd	119.64 ± 1.15 cd	5.23 ± 0.40 b
	10%	11.15 ± 0.24 i	3.36 ± 0.22 b	121.23 ± 1.50 abc	6.20 ± 0.24 a
Date	2.5%	15.19 ± 0.02 c	2.77 ± 0.15 e	112.00 ± 1.18 gh	4.04 ± 0.41 cd
	5%	13.46 ± 0.03 e	2.90 ± 0.14 de	115.10 ± 1.50 ef	4.40 ± 0.55 c
	7.5%	12.27 ± 0.05 h	3.10 ± 0.22 cd	118.23 ± 1.30 de	5.15 ± 0.32 b
	10%	10.18 ± 0.02 k	3.23 ± 0.13 bc	122.10 ± 1.52 ab	6.71 ± 0.17 a
Control	0%	17.30 ± 0.01 a	2.54 ± 0.01 f	98.50 ± 1.50 i	3.80 ± 0.22 d
F		**3181.54**	**21.63**	**58.27**	**33.46**

Means in the same column with different letters are significantly different ($p < 0.05$): DM: dry matter.

Concerning the OCT, Table 2 reveals that the substitution of wheat flour with by-products significantly reduced OCT from 17.30 min (control) to ~10.18 min (10% by-products). Furthermore, the increase in the by-product rate from 2.5% to 10% considerably reduced the OCT from 16.20 min to 10.18 min, respectively. These results are in concordance with those obtained by Kuchtová et al. [28], Petitot et al. [29], and Bouacida et al. [16] after the addition of wheat flour to pasta with Vicia faba, pumpkin, bean flour, eruca vesicaria leaves, and oat bran powder, respectively. Gluten is a major ingredient responsible for the development of the starch–protein structure. Therefore, a dilution of these components with fiber by-products reduces the OCT, as shown in [1,30]. According to Lucas-Gonzalez et al. [1], reducing the OCT of spaghetti could provide the market with products that require less processing time.

Table 2 shows that the supplementation of by-products can lead to higher swelling and CWA of the pasta (Table 2). A great rate of substitution with by-products is associated with a very high swelling capacity and water absorption, reflecting the effect of the fiber on the swelling power of the samples.

Regarding the SI, Table 2 shows that its values ranged from 2.54% (for the control) to 3.86% (for the pasta containing 10% by-products). The greatest SI was obtained for pasta containing 10% by-products. This may be due to the great capacity of fibers to absorb and retain water within a very-well-developed starch–protein–polysaccharide network. The results support those found by Tudorica et al. [4] and Bouacida et al. [16], showing the effect of pea fiber and Eruca vesicaria leaves powder on the CL and SI of pasta.

CWA indicates the amount of water absorbed by the pasta during cooking [31]. According to the MSZ 20500/1–1985 standard, CWA must be at least 100% of the dry pasta mass. All formulations of pasta achieved the minimum requirement. It was shown that the addition of by-products from 2.5% to 10% gradually increased CWA from 98.0% (control) to 123.2% (10%). The pasta with the highest quantity of by-products showed the highest CWA. These observations can be explained by the competition between the starch and fiber for the absorption of water [30]. These results are consistent with those obtained by Rosa-Sibakov et al. [32] (faba starch), Kuchtová et al. [28] (pumpkin powder), and Aravind et al. [25] (guar gum and carboxymethylcellulose).

CL is one of the most important parameters that can predict the overall pasta cooking performance by both industry and consumers, with a low value showing good quality [33]. Results revealed that the CL for all fiber-enriched samples was slightly higher than that of the control (3.80%). According to Table 2, the CL increased from 4.04% to 6.71% as the rate of by-products was increased from 2.5% to 10%. These values are close to those obtained by

Wang et al. [34], showing an enhancement of CL from 4.95% to 7.57% when the rate of rice bran fiber was ~15%. The CL values are similar to those observed for pasta supplemented with pea fibers (5.77% < CL < 6.99%) [4]. By-product-enriched pastas are judged to be of excellent quality because the CL was smaller than 12% for all rates of addition [35]. The results agree with those obtained by Kaur et al. [36], who exposed a positive linear correlation between the CL and the rate of bran addition. The increase in CL may be due to a disruption in the protein–starch matrix and the uneven distribution of water within the pasta matrix due to the competitive hydration tendency of fiber by-products. This explanation agrees with the results reported by Tudorica et al. [4], who exposed the interactions of pea fibers in pasta products. Therefore, the presence of by-products disrupts the protein network and increases the CL.

3.2.2. Texture Analysis

The mean values of adhesiveness and firmness for cooked pasta are summarized in Table 3. Statistical analysis showed that all textural parameters were affected by the addition of by-products ($p < 0.05$) as well as during storage. In addition, as shown in Table 3, pastas with enriched apple, pear, and date by-products have similar textural properties. The increase in by-product supplementation caused a decrease in firmness (from 12.50 N to 7.24 N) and an increase in the adhesiveness (from −0.96 N.s to −0.17 N.s) compared with the formulation with no fiber by-products. The results are in line with those exposed by Silva et al. [37], showing that the additional rates of 5 and 10% with barley and oat bran formulations seemed to cause a decrease in pasta firmness. Texture analysis shows that the firmness of pasta supplemented with by-products was lower than that of the control (Table 3). This would suggest that by-products destabilize the structure strength of pasta. This hypothesis could also be related to the high values obtained for CL, showing a destroyed structure from which large rates of solids are produced during cooking. This event was exposed by Wojtowicz and Moscicki [38] as well as Bouacida et al. [16]. Bustos et al. [6] showed that the extra oat bran disrupted the formation of the protein–starch matrix of the pasta, leading to lower values of firmness. The decrease in pasta firmness could be associated with the role of non-gluten proteins or the insoluble fiber present in the plant material (by-products), which might interfere with the continuity of the gluten matrix, probably making it weaker [25,39].

The pastas enriched with by-products were less sticky than the control. A similar trend was appreciated in vermicelli adhesiveness values after the addition of 20 g/100 g of wheat bran [40] as well as in fresh pasta supplemented with inulin and pea fiber [41]. Statistical analysis showed that while firmness decreased adhesiveness increased significantly ($p < 0.05$) during conservation (Table 2). Similar observations are reported by Borneo and Aguirre [5] as well as Bouacida et al. [16]. This may be due to the high amount of fiber in by-products, which influenced the texture of pasta.

3.2.3. Color Characteristics of Pasta

CIELAB coordinates (L*a*b*) of all pastas are indicated in Table 4. The L*, a*, and b* values of the control (0% by-product) were 70.20 ± 2.23, 5.93 ± 0.15, and 25.54 ± 1.25, respectively. These results showed that the control sample has higher L*, a*, and b* values. Islas-Rubio et al. [39] showed that pasta products made from semolina with high L* and b* values induce a more desirable product. Similar observations were obtained by Bouacida et al. [16]. Indeed, pasta enriched with by-products induced a decrease in all color parameters from 70.20 ± 2.23, 5.93 ± 0.15, and 25.54 ± 1.25 to ~41, ~−1, and ~8 for L*, a*, and b*, respectively. This is in concordance with the findings of Abdel-Moemin et al. [7]. By-products have an apparent effect on the a* value, leading to a greener hue in the pasta. This enhancement in greenness could be due to the lignin component that composes fiber; lignin has an aromatic structure and might cause more Maillard reactions with other components in the food matrix [6]. Therefore, the pastas enriched with by-products have a darker tint. The results obtained in the present study are in concordance with data mentioned in the literature; specifically, pasta enriched with dietary fiber was significantly darker (lower L*) [42].

In addition, results showed that an increase in the rate of by-products causes cooked pasta to be darker (decrease in L* value) (Table 4). CIELAB coordinates 'b*' (positive values) and 'a*' (negative values) (Table 2) significantly decreased due to pasta ingredients, essentially the color of the by-products, and were mainly associated with caramelization and Maillard reactions during the preparation of cooked pasta. In addition, browning reactions could also be due to the deterioration of unsaturated galacturonides in pectin that happened during the drying step [42].

Results did not reveal a significant difference between the samples according to the storage time for all color parameters. This reveals that the color of pasta was stable during all periods of conservation. Comparable observations were reported by Cemin et al. [43] for pasta supplemented with spinach and broccoli.

Furthermore, the color difference (ΔE) increased significantly ($p < 0.05$) as the rate of by-products was increased from 2.5% to 10%, which agrees with Barbara et al. [2]. According to the color difference, the results of pasta revealed an observable difference (ΔE > 5) amongst all samples. This shows the presence of two different colors [44]. The color changes of the dry pasta during storage could be due to enzymatic processes and non-enzymatic oxidation [2].

Generally, it could be concluded that all fiber source materials altered the L*, a*, and b* values, resulting in all formulations having a more brownish hue compared to the reference (control).

Table 3. Firmness and adhesiveness of cooked pasta made with added apple, pear, and date by-products.

By-Product Addition (g/100 g DM)			T_0	T_1	T_2	T_3	*F*
Apple	*Firmness (N)*	2.5%	8.25 ± 0.20 [a]	8.00 ± 0.11 [ab]	7.55 ± 0.05 [b]	6.93 ± 0.05 [c]	**29.88**
		5%	8.05 ± 0.10 [a]	7.25 ± 0.12 [ab]	6.86 ± 0.05 [b]	6.04 ± 0.01 [b]	**4.263**
		7.5%	7.65 ± 0.21 [a]	6.35 ± 0.05 [b]	6.02 ± 0.21 [c]	5.85 ± 0.15 [c]	**95.88**
		10%	7.25 ± 0.13 [a]	6.21 ± 0.10 [b]	5.86 ± 0.15 [c]	5.10 ± 0.12 [d]	**154.6**
	Adhesiveness (N.s)	2.5%	−0.24 ± 0.01 [a]	−0.22 ± 0.02 [a]	−0.20 ± 0.01 [a]	−0.19 ± 0.01 [a]	**0.55**
		5%	−0.22 ± 0.00 [c]	−0.19 ± 0.01 [b]	−0.16 ± 0.02 [a]	−0.14 ± 0.01 [a]	**24.50**
		7.5%	−0.20 ± 0.02 [c]	−0.15 ± 0.01 [b]	−0.13 ± 0.01 [ab]	−0.11 ± 0.02 [a]	**17.90**
		10%	−0.18 ± 0.01 [c]	−0.14 ± 0.01 [b]	−0.12 ± 0.02 [ab]	−0.10 ± 0.02 [a]	**14.00**
Pear	*Firmness (N)*	2.5%	8.46 ± 0.15 [a]	7.22 ± 0.02 [b]	6.85 ± 0.10 [c]	6.51 ± 0.18 [d]	**132.13**
		5%	8.26 ± 0.20 [a]	7.13 ± 0.15 [b]	6.64 ± 0.01 [c]	6.05 ± 0.15 [c]	**30.86**
		7.5%	8.05 ± 0.10 [a]	7.02 ± 0.10 [b]	6.45 ± 0.16 [c]	5.85 ± 0.10 [d]	**192.28**
		10%	7.55 ± 0.50 [a]	6.95 ± 0.30 [b]	6.01 ± 0.12 [c]	5.65 ± 0.10 [c]	**63.53**
	Adhesiveness (N.s)	2.5%	−0.25 ± 0.01 [b]	−0.21 ± 0.05 [ab]	−0.20 ± 0.01 [ab]	−0.17 ± 0.01 [a]	**5.17**
		5%	−0.23 ± 0.01 [c]	−0.19 ± 0.03 [b]	−0.17 ± 0.01 [ab]	−0.15 ± 0.02 [a]	**9.33**
		7.5%	−0.21 ± 0.01 [b]	−0.20 ± 0.01 [b]	−0.18 ± 0.01 [b]	−0.13 ± 0.05 [a]	**5.42**
		10%	−0.20 ± 0.02 [c]	−0.18 ± 0.02 [bc]	−0.15 ± 0.02 [ab]	−0.14 ± 0.01 [a]	**7.00**
Date	*Firmness (N)*	2.5%	7.95 ± 0.20 [a]	7.04 ± 0.35 [b]	6.75 ± 0.15 [c]	5.96 ± 0.11 [d]	**69.04**
		5%	7.85 ± 0.30 [a]	6.98 ± 0.06 [b]	6.45 ± 0.05 [c]	5.76 ± 0.10 [d]	**99.91**
		7.5%	7.55 ± 0.20 [a]	6.45 ± 0.15 [b]	5.85 ± 0.11 [c]	5.15 ± 0.05 [d]	**142.61**
		10%	7.24 ± 0.18 [a]	6.19 ± 0.20 [b]	5.64 ± 0.12 [c]	5.01 ± 0.10 [d]	**103.50**
	Adhesiveness (N.s)	2.5%	−0.20 ± 0.03 [b]	−0.17 ± 0.01 [ab]	−0.15 ± 0.02 [a]	−0.13 ± 0.02 [a]	**5.94**
		5%	−0.18 ± 0.01 [b]	−0.17 ± 0.01 [b]	−0.16 ± 0.02 [ab]	−0.14 ± 0.01 [a]	**5.00**
		7.5%	−0.17 ± 0.01 [c]	−0.15 ± 0.01 [bc]	−0.14 ± 0.01 [ab]	−0.12 ± 0.02 [a]	**7.42**
		10%	−0.17 ± 0.01 [b]	−0.14 ± 0.02 [a]	−0.12 ± 0.01 [a]	−0.11 ± 0.02 [a]	**8.40**
Control	*Firmness(N)*	0%	12.50 ± 0.05 [a]	10.25 ± 0.15 [b]	9.55 ± 0.10 [c]	9.02 ± 0.01 [d]	**802.25**
	Adhesiveness (N.s)		−0.96 ± 0.01 [c]	−0.85 ± 0.03 [b]	−0.80 ± 0.03 [a]	−0.76 ± 0.01 [a]	**28.37**

T_0: 0 days; T_1: 7 days; T_2: 21 days; and T_3: 30 days of storage. Mean (n = 3) ± SD. Means in the same line with different letters are significantly different ($p < 0.05$). DM: dry matter.

Table 4. Color parameters of enriched pasta during storage.

		Pear				Apple				Date			
		T_0	T_1	T_2	T_3	T_0	T_1	T_2	T_3	T_0	T_1	T_2	T_3
L*	2.5%	56.24 ±1.72	54.20 ±0.13	54.01 ±0.41	53.01 ±2.01	53.17 ±0.15	53.69 ±0.31	52.25 ±0.21	52.01 ±0.11	53.04 ±0.38	53.65 ±0.56	52.72 ±0.14	52.01 ±0.04
	5%	55.34 ±0.51	53.30 ±1.41	49.44 ±1.20	49.01 ±0.27	53.22 ±1.50	53.70 ±2.01	53.60 ±1.03	52.20 ±1.23	45.79 ±0.46	43.32 ±0.57	44.72 ±0.65	41.25 ±1.75
	7.5%	49.48 ±1.01	47.78 ±2.12	47.21 ±0.01	46.00 ±1.25	52.67 ±0.06	52.55 ±0.75	51.44 ±0.38	50.62 ±0.34	41.61 ±0.08	40.26 ±0.04	40.76 ±0.31	38.54 ±0.02
	10%	41.38 ±1.51	40.23 ±1.26	40.33 ±2.01	38.56 ±1.14	51.86 ±0.10	50.73 ±0.15	50.01 ±0.01	48.41 ±2.02	39.29 ±0.45	38.84 ±0.08	37.20 ±0.75	36.75 ±0.13
a*	2.5%	4.87 ±0.18	4.60 ±0.06	4.20 ±0.05	4.16 ±0.02	4.27 ±0.45	4.15 ±0.23	4.11 ±0.06	4.12 ±0.02	3.56 ±0.10	3.16 ±0.05	3.17 ±0.01	3.20 ±0.38
	5%	1.53 ±0.03	1.50 ±0.01	1.49 ±0.04	1.42 ±0.02	1.64 ±0.02	1.31 ±0.18	1.11 ±0.07	1.10 ±0.13	1.89 ±0.25	1.56 ±0.36	1.16 ±0.31	1.11 ±0.01
	7.5%	0.46 ±0.02	0.45 ±0.04	0.43 ±0.07	0.44 ±0.01	0.41 ±0.03	0.40 ±0.01	0.40 ±0.04	0.38 ±0.01	0.75 ±0.05	0.71 ±0.02	0.72 ±0.06	0.70 ±0.05
	10%	−1.40 ±0.67	−1.38 ±0.29	−1.39 ±0.38	−1.37 ±0.13	−1.72 ±0.14	−1.70 ±0.04	−1.69 ±0.01	−1.67 ±0.06	−1.56 ±0.02	−1.57 ±0.01	−1.55 ±0.06	−1.52 ±0.08
b*	2.5%	16.56 ±0.05	16.33 ±0.18	16.28 ±0.03	16.13 ±0.02	18.40 ±0.10	18.14 ±0.01	18.10 ±0.02	18.16 ±0.02	12.78 ±0.05	12.24 ±0.31	12.15 ±0.14	12.10 ±0.01
	5%	13.57 ±0.36	13.36 ±0.76	13.20 ±0.29	13.00 ±0.12	16.32 ±0.07	16.04 ±0.01	15.97 ±0.05	16.00 ±0.02	10.83 ±0.04	10.17 ±0.03	10.11 ±0.20	10.02 ±0.13
	7.5%	10.45 ±0.46	10.25 ±0.11	10.23 ±0.03	10.13 ±0.02	14.70 ±0.03	14.15 ±0.11	14.05 ±0.06	14.00 ±0.02	8.36 ±0.01	8.08 ±0.04	7.97 ±0.01	8.00 ±0.06
	10%	8.59 ±0.22	8.23 ±0.01	8.20 ±0.14	8.25 ±0.24	12.83 ±0.05	12.26 ±0.05	12.13 ±0.01	12.10 ±0.02	5.19 ±0.01	5.10 ±0.31	5.06 ±0.14	5.08 ±0.05
ΔE	2.5%	14.14 ±0.03	15.88 ±0.01	16.05 ±0.25	16.96 ±0.15	10.69 ±0.10	15.34 ±0.25	16.63 ±0.21	16.82 ±0.14	22.13 ±0.25	18.95 ±0.10	24.41 ±0.05	20.24 ±0.01
	5%	16.96 ±0.20	18.58 ±0.05	21.70 ±0.10	22.17 ±0.52	13.72 ±0.21	16.65 ±0.18	16.80 ±0.34	17.93 ±0.19	25.99 ±0.23	28.43 ±0.14	27.34 ±0.27	30.30 ±0.05
	7.5%	23.46 ±0.15	24.90 ±0.10	26.96 ±0.23	27.18 ±0.14	16.16 ±0.14	18.83 ±0.28	19.76 ±0.26	20.45 ±0.31	30.92 ±0.25	32.20 ±0.05	31.84 ±0.20	33.69 ±0.34
	10%	31.28 ±0.18	32.42 ±0.30	32.35 ±0.05	33.80 ±0.10	20.14 ±0.15	21.83 ±0.27	22.45 ±0.04	23.71 ±0.16	34.12 ±0.14	35.32 ±0.34	36.67 ±0.25	37.02 ±0.13

T_0: 0 days; T_1: 7 days; T_2: 21 days; and T_3: 30 days of storage. Mean (n = 3) ± SD.

3.2.4. Water Activity Characteristics of Dried Pasta

Table 1 shows that the a_w measurements of all by-product-supplemented pasta samples were overall higher than that of the control. This tendency could be due to higher water absorption by fibers. Indeed, several authors argue that the dietary fibers supplemented into food products can enhance oil and water holding capacity, emulsification, and/or gel formation [45,46]. Indeed, dietary fiber structures present an important number of hydroxyl groups which provide more water interactions through hydrogen bonding [47]. Table 5 shows that enriched pastas have similar a_w values. This could be due to the similar chemical composition of all fibers from cooked fruit by-products [15]. In addition, Table 5 shows that a_w increased as the rate of fibers increased from 2.5% to 10%. Results obtained in this study are parallel to those reported by Sudha et al. [40] at 5, 10, and 15 percent substitution rates of oat bran in pasta. After 30 days of storage all formulated pastas had noticeably lower a_w rates, ranging from 0.537 to 0.570, compared to the beginning of the process (0.701 < aw < 0.672). This fact is due to water loss during the conservation process. Moreover, statistical analyses revealed a significant difference ($p < 0.05$) between all formulated pastas during storage. The water activity (a_w) values observed in the current work are in accordance with those reported by Aramouni and Mahmoud [48], Sun-Waterhouse and Wadhwa, [46] and Silva et al. [37], showing that the a_w of pasta varied between 0.226–0.650.

Table 5. Water activity of by-product-enriched pasta formulations during storage.

By-Product Addition (g/100 g DM)		T_0	T_1	T_2	T_3	*F*
Apple	2.5%	0.688 ± 0.005 ^{a}	0.585 ± 0.002 ^{b}	0.573 ± 0.005 ^{c}	0.565 ± 0.004 ^{d}	**602.20**
	5%	0.686 ± 0.001 ^{a}	0.592 ± 0.005 ^{b}	0.577 ± 0.001 ^{c}	0.568 ± 0.003 ^{d}	**986.75**
	7.5%	0.685 ± 0.004 ^{a}	0.597 ± 0.003 ^{b}	0.580 ± 0.003 ^{c}	0.570 ± 0.005 ^{d}	**561.22**
	10%	0.701 ± 0.003 ^{a}	0.599 ± 0.002 ^{b}	0.582 ± 0.005 ^{c}	0.571 ± 0.005 ^{d}	**677.12**
Pear	2.5%	0.695 ± 0.005 ^{a}	0.577 ± 0.006 ^{b}	0.568 ± 0.005 ^{b}	0.546 ± 0.003 ^{c}	**528.18**
	5%	0.699 ± 0.002 ^{a}	0.579 ± 0.003 ^{b}	0.569 ± 0.002 ^{c}	0.549 ± 0.001 ^{d}	**3066.66**
	7.5%	0.701 ± 0.005 ^{a}	0.581 ± 0.002 ^{b}	0.572 ± 0.005 ^{bc}	0.550 ± 0.004 ^{c}	**9.80**
	10%	0.702 ± 0.002 ^{a}	0.583 ± 0.001 ^{b}	0.574 ± 0.005 ^{c}	0.555 ± 0.001 ^{d}	**1721.93**
Date	2.5%	0.689 ± 0.003 ^{a}	0.568 ± 0.001 ^{b}	0.545 ± 0.005 ^{c}	0.540 ± 0.004 ^{c}	**115.21**
	5%	0.691 ± 0.005 ^{a}	0.690 ± 0.001 ^{a}	0.573 ± 0.004 ^{b}	0.544 ± 0.002 ^{c}	**1551.73**
	7.5%	0.698 ± 0.005 ^{a}	0.590 ± 0.006 ^{b}	0.576 ± 0.002 ^{b}	0.551 ± 0.001 ^{c}	**204.59**
	10%	0.700 ± 0.003 ^{a}	0.598 ± 0.005 ^{b}	0.579 ± 0.001 ^{b}	0.555 ± 0.002 ^{c}	**239.85**
Control	0%	0.672 ± 0.001 ^{a}	0.558 ± 0.001 ^{b}	0.541 ± 0.005 ^{b}	0.537 ± 0.003 ^{b}	**16.24**

T_0: 0 days; T_1: 7 days; T_2: 21 days; and T_3: 30 days of storage. Mean (n = 3) ± SD. Means in the same line with different letters are significantly different ($p < 0.05$).

Table 5 shows that the water activity of pasta enriched with dietary fibers remained higher than that of the control after the storage time. The a_w values show that all the dried pastas produced in this study would have a good shelf life. In fact, all a_w rates are below 0.650; thus, enriched pasta should be stable against microbial growth and might persist for about six months [48].

3.2.5. Rheological Properties of Enriched Pasta

The effect of apple, pear, and date by-product addition at different rates on the alveograph parameters of pasta dough are shown in Table 6. The addition of by-products always significantly ($p < 0.05$) enhances dough tenacity (P) (from 80.23 to ~135 mm of H_2O) and significantly reduces dough extensibility (L) (from 62.53 to ~17 mm) compared to the control. Therefore, doughs have the ability to retain gas with a low capacity of extending without breaking down. The enhancement of P values is likely due to the interaction between flour protein and fiber structure, or the bad hydration of doughs supplemented with by-products [34]. In fact, by-product-enriched pastas require more water than the control due to their significant number of hydroxyl groups which allow more water interactions. The determination of the value of the P/L ratio gives information about the extensibility and elastic resistance balance of flour dough and summarizes the effect of L and P parameters. Table 6 shows that the value of the P/L ratio increased (from 1.29 to 7.72) as the rate of fibers increased from 2.5% to 10%. Wang et al. [34] claim that an increase in the P/L parameter might be caused by the high content of cellulose present in the fibers, which promotes a powerful interaction between fibers and flour protein. Table 6 shows a statistical difference ($p < 0.05$) between the enriched pasta and the control concerning the deformation energy. In fact, the dough deformation energy decreased significantly ($p < 0.05$) (from 154×10^{-4} to ~124×10^{-4}) with the addition of by-products. Similar observations are reported by Shreenithee and Prabhasankar, [27] revealing that the addition of yellow pea flour reduces the dough deformation energy.

3.2.6. Thermal Gelling Properties of Pasta

The impacts of the inclusion of apple, date, and pear by-products on starch gelatinization properties were investigated using DSC methodology. Results from DSC showed that the addition of by-products affects the thermal gelling properties of pasta (Table 3). The starch gelatinization temperature of enriched pasta decreased proportionally as the rate of by-products was raised from 2.5% to 10%. As such, the results are consistent with previous research [4], which reveals that the addition of soluble non-starch polysaccharides reduces gelatinization temperature. This is partly due to the soluble fibers competing with starch for water absorption and therefore limiting gelatinization as well as starch swelling events, resulting in a lower than expected T_{endset} value (Table 7). Martín-Esparza et al. [49] revealed a similar result. If fact, the authors found that incorporating tiger nut flour into pasta reduced the starch gelatinization temperature due to the presence of a smaller amount of starch available for gelatinization; therefore, a lower amount of energy is needed for such a transition to occur.

Table 6. Alveograph characteristics of dough containing different rates and kinds of fiber by-products.

		Tenacity (P) (mm of H_2O)	Extensibility (L) (mm)	P/L	Deformation Energy ($\times 10^{-4}$)
Dough Control	0%	80.23 ± 1.80^{h}	62.53 ± 0.40^{a}	1.29 ± 0.05^{g}	154.82 ± 2.30^{a}
Flour with Date Fibers	2.5%	115.23 ± 1.50^{g}	25.84 ± 0.50^{b}	4.46 ± 0.21^{f}	135.21 ± 1.15^{b}
	5%	123.16 ± 1.50^{ef}	23.64 ± 1.20^{bcd}	5.21 ± 0.03^{e}	133.11 ± 2.23^{bc}
	7.5%	130.52 ± 2.30^{bc}	20.15 ± 1.40^{defg}	6.47 ± 0.01^{c}	130.50 ± 1.01^{cde}
	10%	135.43 ± 1.20^{a}	18.01 ± 1.50^{fg}	7.52 ± 0.22^{a}	128.30 ± 2.84^{ef}
Flour with Pear Fibers	2.5%	114.12 ± 1.40^{g}	26.23 ± 1.50^{b}	4.35 ± 0.10^{f}	$133.10 \pm 1.5\ 0^{bcd}$
	5%	122.40 ± 2.80^{ef}	23.60 ± 3.70^{bcd}	5.18 ± 0.13^{e}	129.50 ± 5.30^{def}
	7.5%	128.00 ± 1.50^{cd}	21.54 ± 1.10^{cdef}	5.94 ± 0.23^{d}	127.20 ± 2.10^{efg}
	10%	133.60 ± 3.50^{ab}	18.76 ± 3.80^{fg}	7.12 ± 0.01^{b}	125.23 ± 1.40^{fg}
Flour with Apple Fibers	2.5%	121.23 ± 3.50^{f}	24.13 ± 1.40^{bc}	5.02 ± 0.11^{e}	134.23 ± 2.80^{bc}
	5%	126.30 ± 2.10^{de}	22.10 ± 1.70^{cde}	5.71 ± 0.02^{d}	130.50 ± 3.50^{bc}
	7.5%	130.23 ± 1.80^{bc}	19.50 ± 1.20^{efg}	6.68 ± 0.22^{c}	126.23 ± 2.40^{efg}
	10%	135.10 ± 2.70^{a}	17.50 ± 1.90^{g}	7.72 ± 0.01^{a}	124.10 ± 1.50^{g}
F		**142.84**	**118.49**	**531.40**	**34.47**

Means in the same column with different letters are significantly different ($p < 0.05$).

Table 7. Thermal gelling properties of pasta dough.

	By-Product Addition (g/100 g DM)	T_{onset} (°C)	T_{endset} (°C)	Enthalpy (ΔH j/g)	Gelatinization Temperature (°C)
Control	0%	57.23 ± 0.25^{a}	67.42 ± 0.29^{c}	0.97 ± 0.06^{a}	62.07 ± 0.22^{bc}
Pear	2.5%	56.03 ± 0.40^{b}	67.89 ± 0.25^{bc}	0.80 ± 0.01^{bcd}	62.31 ± 0.16^{c}
	5%	55.56 ± 0.45^{bc}	68.06 ± 0.35^{bc}	0.71 ± 0.03^{ef}	62.54 ± 0.05^{bc}
	7.5%	55.30 ± 0.30^{cd}	68.35 ± 0.15^{abc}	0.66 ± 0.03^{fg}	62.61 ± 0.05^{abc}
	10%	55.05 ± 0.21^{cd}	68.20 ± 0.65^{bc}	0.62 ± 0.01^{g}	62.84 ± 0.20^{a}
Apple	2.5%	56.88 ± 0.43^{a}	68.01 ± 0.12^{bc}	0.84 ± 0.02^{b}	61.10 ± 0.35^{d}
	5%	56.01 ± 0.31^{b}	68.25 ± 0.35^{abc}	0.76 ± 0.01^{cde}	62.29 ± 0.12^{c}
	7.5%	55.50 ± 0.26^{bc}	68.45 ± 0.10^{c}	0.68 ± 0.02^{de}	62.36 ± 0.23^{c}
	10%	55.25 ± 0.29^{cd}	68.81 ± 0.20^{a}	0.65 ± 0.03^{fg}	62.44 ± 0.25^{bc}
Date	2.5%	55.86 ± 0.32^{b}	67.95 ± 0.14^{bc}	0.82 ± 0.01^{bc}	62.37 ± 0.30^{c}
	5%	55.10 ± 0.15^{cd}	68.08 ± 0.23^{bc}	0.74 ± 0.02^{cde}	62.56 ± 0.15^{abc}
	7.5%	54.89 ± 0.10^{d}	68.55 ± 0.15^{ab}	0.64 ± 0.02^{g}	62.78 ± 0.11^{abc}
	10%	54.75 ± 0.19^{d}	68.70 ± 0.22^{ab}	0.60 ± 0.01^{g}	62.93 ± 0.20^{ab}
	F	**19.10**	**2.75**	**21.95**	**9.08**

Means in the same column with different letters are significantly different ($p < 0.05$).

Table 7 shows that the enthalpy of an enriched sample decreased with an increase in the rate of fibers. This rise was not moderate and showed significant ($p < 0.05$) differences between the control and fibers from different origins. The enthalpy of a system is an indicator of the quantity of starch gelatinization within a flour or starch base, and should therefore be linked to the gelatinization temperature of starch. One possible explanation for the pattern of enthalpy values of fiber systems is that, at a high rate of by-product, the fiber greatly competes for water with starch, affecting gelling and pasting events. Therefore, the addition of apple, pear, and date by-products to pasta has a significant ($p < 0.05$) impact on starch gelatinization properties.

3.3. Scanning Electron Microscopy of Pasta Dough

The internal structure of pasta dough is shown in Figure 1. The micrograph of the control dough sample shows that the protein–starch matrix is well-formed, with continuous and strong protein strands entrapping large starch granules. In fact, the control sample presents swollen and gelatinized starch granules which appear to be integrated into a developed protein matrix to make a compact structure, with some starch granules not gelatinized. These results are in accordance with those of Bustos et al. [6].

Micrographs of pasta dough with the inclusion of by-products show a similar protein–fiber matrix for all types of by-product-enriched pastas. The supplementation of by-products into the pasta involves a disruption of the continuity of the protein matrix. The protein–fiber matrix within pastas containing by-products at 5%, 7.5%, and 10% appears to be less developed than that of the control, resulting in an open appearance with discrete starch granules 'uncovered' and exposed to enzymatic attack [6]. The degree of disruption appears to increase with the rate of fibers added to the product. However, the addition of by-products at 2.5% appears not to significantly affect pasta structure compared with that of the control. These results are in accordance with those obtained by replaced Ainsa et al. [19] and Bustos et al. [6]. This disruption to the pasta structure may explain the decrease in firmness observed in Table 3. A modified pasta microstructure can affect the rate of starch degradation which affects the insulinemic and glycemic indices [6].

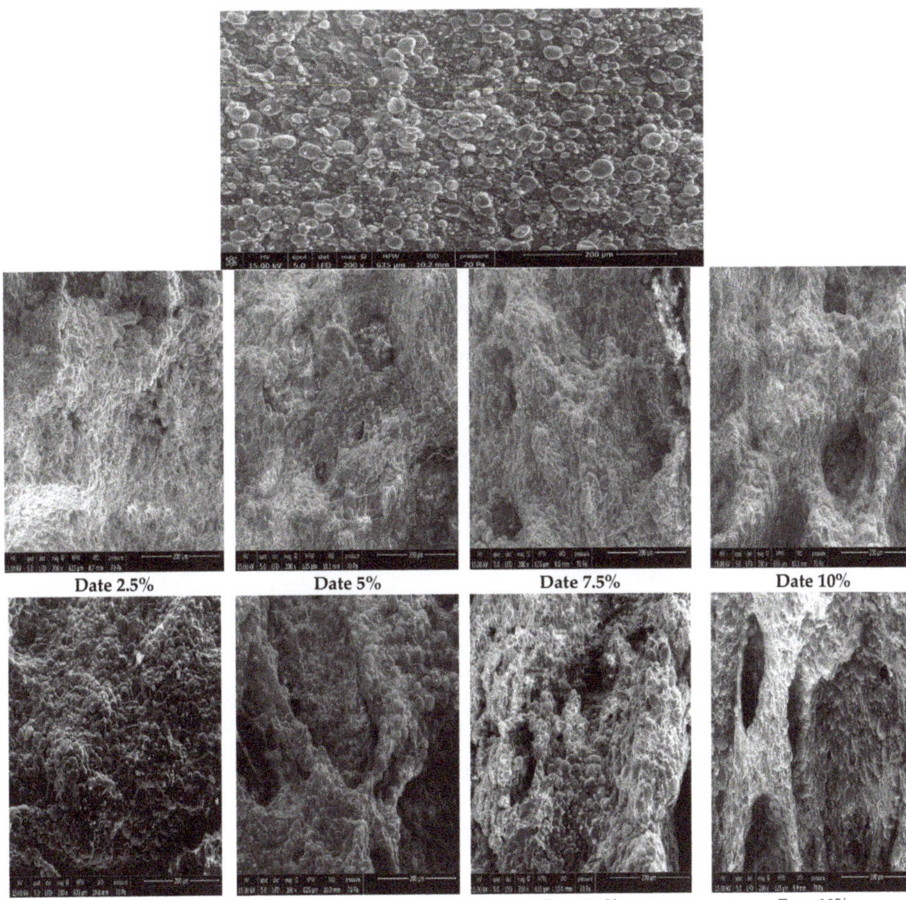

Figure 1. *Cont.*

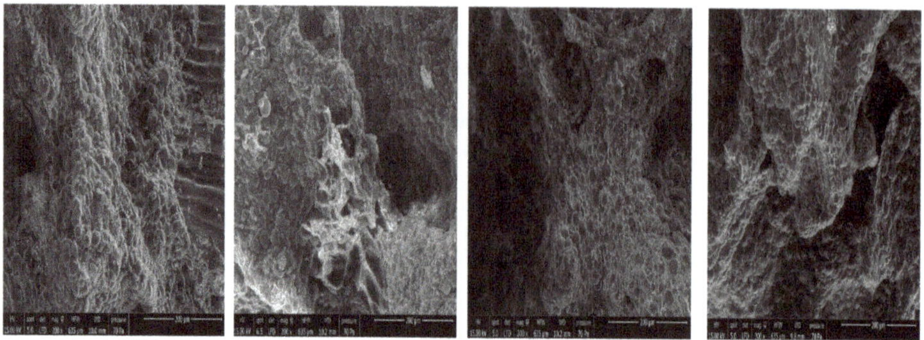

Apple 2.5% Apple 5% Apple 7.5% Apple 10%

Figure 1. Scanning electron microscopy of dough enriched with date, apple, and pear by-products.

3.4. Sensory Evaluation

The sensory properties of cooked pasta enriched with apple, pear, and date by-products were evaluated to choose the best-suited substitute flour for the formulation of pasta. The mean scores of sensory attributes are shown in Table 8. Statistical analysis did not show a significant difference ($p > 0.05$) for taste scores between all pasta formulations. Therefore, consumers could not differentiate the taste of pasta between different percentages of substitution.

Table 8. Sensory attributes of cooked pasta.

		Taste	Aftertaste	Appearance	Texture	Color	Overall Acceptability
Control	**0%**	6.5 ± 0.4 ^{a}	6.6 ± 0.5 ^{c}	6.0 ± 0.3 ^{bcd}	6.3 ± 0.2 ^{ab}	5.9 ± 0.4 ^{bc}	6.2 ± 0.2 ^{abc}
Flour with Apple Fibers	**2.5%**	6.6 ± 0.1 ^{a}	6.5 ± 0.2 ^{ab}	6.2 ± 0.1 ^{ab}	6.5 ± 0.3 ^{a}	6.4 ± 0.1 ^{a}	6.4 ± 0.1 ^{a}
	5%	6.5 ± 0.3 ^{a}	6.0 ± 0.1 ^{c}	6.0 ± 0.2 ^{bcd}	6.0 ± 0.1 ^{bcd}	5.9 ± 0.1 ^{b}	6.0 ± 0.1 ^{cde}
	7.5%	6.3 ± 0.2 ^{ab}	6.1 ± 0.2 ^{bc}	5.9 ± 0.3 ^{be}	5.4 ± 0.4 ^{fg}	5.6 ± 0.2 ^{cde}	5.8 ± 0.2 ^{ef}
	10%	6.4 ± 0.1 ^{ab}	6.0 ± 0.1 ^{c}	5.8 ± 0.1 ^{cde}	5.2 ± 0.2 ^{g}	5.4 ± 0.2 ^{e}	5.7 ± 0.1 ^{f}
Flour with Pear Fibers	**2.5%**	6.4 ± 0.1 ^{ab}	6.5 ± 0.2 ^{ab}	6.1 ± 0.1 ^{abc}	6.2 ± 0.2 ^{abc}	6.5 ± 0.1 ^{a}	6.3 ± 0.1 ^{ab}
	5%	6.2 ± 0.1 ^{abc}	6.0 ± 0.2 ^{c}	5.8 ± 0.1 ^{cde}	5.9 ± 0.1 ^{cde}	6.0 ± 0.1 ^{bc}	5.9 ± 0.2 ^{def}
	7.5%	6.0 ± 0.2 ^{bc}	6.1 ± 0.1 ^{bc}	5.6 ± 0.2 ^{e}	5.6 ± 0.1 ^{ef}	5.8 ± 0.1 ^{bcd}	5.8 ± 0.2 ^{ef}
	10%	5.8 ± 0.3 ^{c}	6.0 ± 0.2 ^{c}	5.7 ± 0.1 ^{de}	5.5 ± 0.1 ^{fg}	5.5 ± 0.1 ^{de}	5.8 ± 0.1 ^{ef}
Flour with Date Fibers	**2.5%**	6.3 ± 0.1 ^{ab}	6.5 ± 0.1 ^{ab}	6.4 ± 0.1 ^{a}	6.4 ± 0.2 ^{a}	6.5 ± 0.1 ^{a}	6.4 ± 0.1 ^{a}
	5%	6.2 ± 0.2 ^{abc}	6.4 ± 0.2 ^{abc}	6.2 ± 0.1 ^{ab}	5.7 ± 0.2 ^{def}	6.0 ± 0.2 ^{b}	6.1 ± 0.1 ^{bcd}
	7.5%	6.4 ± 0.1 ^{ab}	6.4 ± 0.1 ^{abc}	6.0 ± 0.2 ^{bcd}	5.6 ± 0.1 ^{ef}	5.7 ± 0.1 ^{bcde}	6.2 ± 0.1 ^{abc}
	10%	6.3 ± 0.1 ^{abc}	6.0 ± 0.3 ^{c}	5.8 ± 0.2 ^{cde}	5.4 ± 0.1 ^{fg}	5.5 ± 0.2 ^{de}	5.8 ± 0.2 ^{ef}
F		**2.60**	**3.66**	**4.84**	**14.05**	**13.81**	**8.67**

Means in the same column with different letters are significantly different ($p < 0.05$).

For other sensory attributes, there is a significant difference between different amounts from 5% of added by-product. Hence, the consumer cannot detect the difference in the aftertaste, appearance, texture, and color of pasta between the control pasta and that enriched with 2.5% of added by-product. Therefore, consumers reacted in different ways to all formulations. These results agree with previous SEM observations, showing that the addition of 2.5% of by-product did not significantly affect the structure of the pasta dough compared to the control.

The pasta enriched with 2.5% of by-product presented the highest overall acceptability score. Table 8 indicates that the overall acceptability decreased significantly ($p < 0.05$) as the amount of by-product was increased from 5% to 10%. Therefore, consumers did not appreciate pasta containing a high by-product rate (above 5%). Statistical analysis revealed a significant ($p < 0.05$) difference in overall acceptability between the pasta supplemented with 2.5% of by-product and the other samples (Table 8). In the same way, Sant'Anna et al. [50] and Crizel et al. [8] revealed that fettuccini pasta with the addition of a higher amount of (75 g/kg, 50 g/kg, and 25 g/kg) grape marc and orange by-product fiber powder resulted in lower acceptability.

The textural and color characteristics of food products have a major role in the final acceptance by consumers. Table 8 shows a significant decrease in texture and color scores with an increase in fiber content in pasta. Indeed, our previous investigations showed that apple, pear, and date by-products have low color parameter (L*, a*, and b*) values, inducing a dark color [15]. Therefore, the original colors of by-products significantly affect the appreciation of the consumer.

Therefore, a practical formulation can be chosen, since a significant difference ($p < 0.05$) was shown between overall acceptability scores. In fact, panels have appreciated the pasta enriched with only 2.5% of by-product more than the other formulations.

3.5. Multiple Factor Analysis (MFA)

Physicochemical and sensorial results were subjected to multiple factor analysis (MFA). The MFA plot of pasta is illustrated in Figure 2A,B, which describes the (A) interrelations among analyzed parameters and the (B) positioning of analyzed pastas in comparison to each other, respectively. The first axis (F1) accounted for 59.16% of the total variance and the second (F2) for 19.83%, accounting for 79.00% of the total variance. Therefore, the first two axes have almost 70% of the variability. From this plot, we can expose that F1 axis was positively correlated (localized in the positive axis of F1) with sensory attributes, optimum cooking time, deformation energy, extensibility, and firmness. On the other hand, the variables taste, swelling index, deformation energy, cooking loss, gelatinization temperature, and textural parameters (tenacity, extensibility, and firmness), were positively loaded with the second axes (F2). Therefore, textural parameters (extensibility and firmness), taste, and gelatinization temperature were positively correlated with both axes.

According to the results of the MFA, the pasta samples were divided into three well-defined groups: (1) pasta supplemented with 2.5% of apple, pear, and date fiber by-products, (2) control (0% fiber by-products) (3) all the other formulations (Figure 2B). This corroborates our previous observations showing that the addition of by-products induces a modification of the physicochemical properties of the pasta. Indeed, the control sample is the furthest to the bottom right, and the other samples are located at the upper left position. This confirms our observation, showing that supplemented pastas have a different sensory profile compared to the control.

The combination of the results of Figure 2A,B show that sensory attributes characterize the first group (pasta supplemented with 2.5% of by-products). However, textural parameters (extensibility and firmness) characterize the second group (the control).

According to the results of MFA, extensibility and firmness represent two closest variables (RV: 0.9911). In fact, RV coefficients show the relationship between variables (1: high correlation; 0: low correlation). In addition, cooking water absorption–tenacity (RV: 0.9368), adhesiveness–extensibility (RV: 0.9703), and tenacity–extensibility (RV: 0.9385) constitute closest variables.

In addition, the sensory attributes and principally the overall acceptability were grouped on the right side of the plot, and the pasta supplemented with 5%, 7.5%, and 10% of by-products were grouped on the left side of the MFA plot. The visualization of these factors reveals that a higher by-product rate negatively affects the appreciation of the consumer. This validates our previous results which showed that consumers attribute the highest score to pasta fortified with 2.5% of by-products.

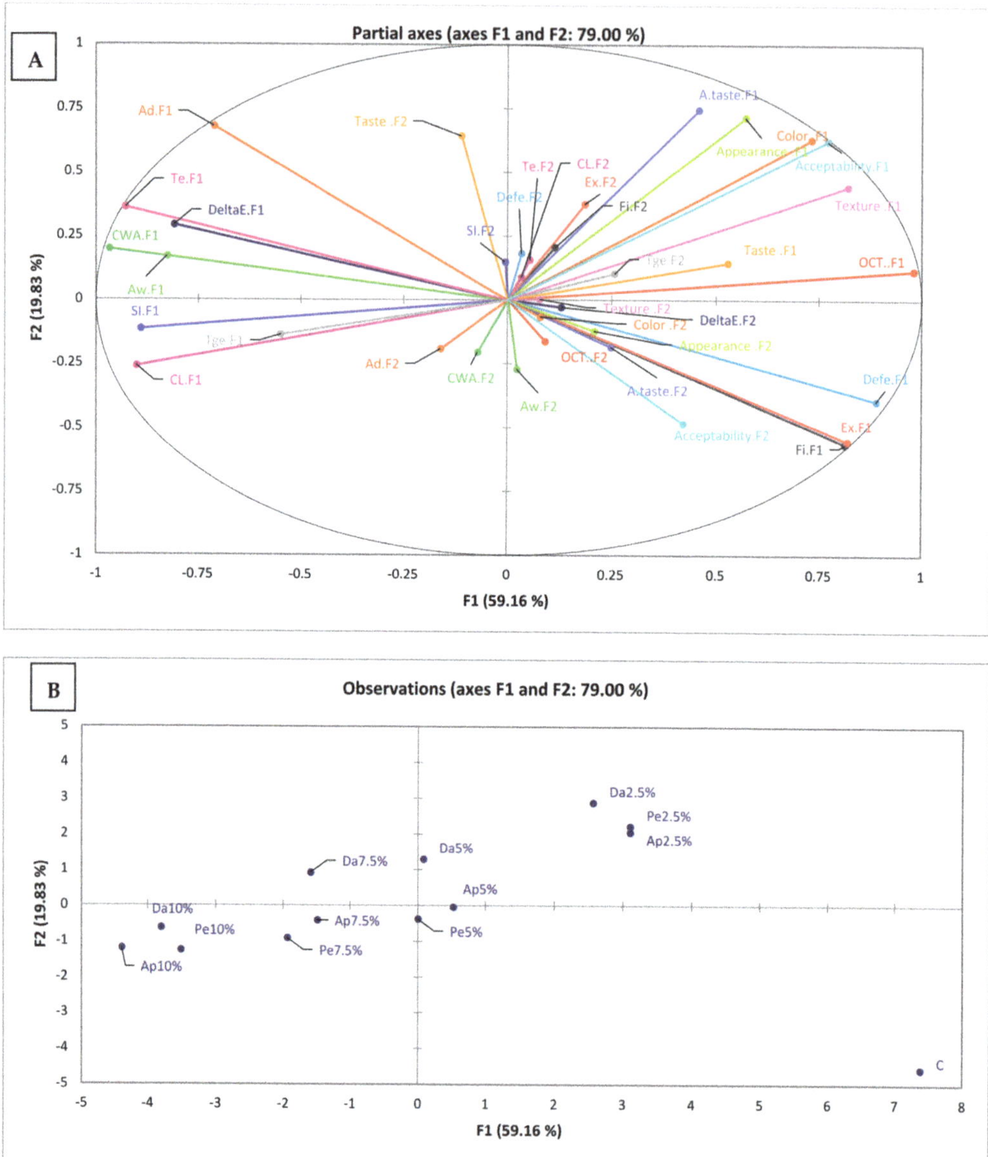

Figure 2. (**A**) The score and loading plots for F1 versus F2 (OCT: optimum cooking time; SI: swelling index; CWA: cooking water absorption; CL: cooking loss; Fi: firmness; Ad: adhesiveness; DeltaE: color difference (ΔE); Aw: water activity; Te: tenacity; Ex: extensibility; Defe: deformation energy; and Tge: gelatinization temperature). (**B**) The distribution of different samples in relation to the F1 and F2 axes.

4. Conclusions

From the overall results, it could be concluded that pasta can be formulated with flour from pear, apple, and date by-products. Indeed, by-products have a positive impact on physicochemical properties and cooking quality attributes of pastas. The addition of by-products improves the fiber content, swelling index, and cooking water absorption of pastas, while reducing their optimum cooking time. However, enhancing the by-product rate (from 2.5% to 10%) negatively alters the texture and the structure of pasta. Indeed, SEM analysis showed that the inclusion of by-products into pasta disrupts the protein matrix. Moreover, the pasta fortified with by-products exhibited a darker color than that of the control semolina pasta. Pear, apple, and date by-products could be added to pasta at a rate of 2.5%. They are considered as the most acceptable organoleptically, with the highest overall appreciation score. This means, that by-products are a suitable material for enriching pasta. In addition, MFA results revealed that the pastas enriched with 2.5% of by-products are distinguished, from the other samples, by their sensory attributes. We also carried out a storage test, which showed a significant modification of the texture and water activity after 30 days, contrary to the color of pasta, which remained stable. Additionally, further research can be carried out on the different shapes of pasta and their effects on the quality of the product as well as the influence of additives in improving the texture of pastas.

Author Contributions: B.B. designed the experiments and wrote the manuscript. S.B. reviewed and edited the manuscript. S.D. gave instructions during experiments. H.A., R.K. and C.B. supervised the project. All authors have read and agreed to the published version of the manuscript.

Funding: This research was funded by the Ministry of Higher Education and Scientific Research, Tunisia.

Institutional Review Board Statement: Not applicable.

Informed Consent Statement: Not applicable.

Data Availability Statement: Not applicable.

Conflicts of Interest: The authors declare no conflict of interest.

References

1. Lucas-Gonzalez, R.; Manuel, V.; Jose, A.; Clemencia, C.; Blerina, S.; Moscaritolo, S.; Juana, F.; Giampiero, S. Persimmon flours as functional ingredients in spaghetti: Chemical, physico-chemical and cooking quality. *J. Food Meas. Charact.* **2020**, *14*, 1634–1644. [CrossRef]
2. Barbara, B.; Rebeka, F.; Ildikó, S.; Klára, P.; Attila, G. Buckwheat-pasta enriched with silkworm powder: Technological analysis and sensory evaluation. *LWT-Food Sci. Technol.* **2019**, *116*, 108542.
3. Minarovičová, L.; Michaela, L.; Zlatica, K.; Jolana, K.; Veronika, K. Effect of pumpkin powder incorporation on cooking and sensory parameters of pasta. *J. Food Sci.* **2017**, *11*, 373–379. [CrossRef]
4. Tudorica, C.M.; Kuri, V.; Brennan, S. Nutritional and Physicochemical Characteristics of Dietary Fiber Enriched Pasta. *J. Agric. Food Chem.* **2002**, *50*, 347–356. [CrossRef]
5. Borneo, R.; Aguirre, A. Chemical composition, cookingquality, and consumer acceptance of pasta made with dried amaranth leaves flour. *LWT-Food Sci. Technol.* **2008**, *41*, 1748–1751. [CrossRef]
6. Bustos, M.C.; Pérez, G.T.; León, A.E. Effect of four types of dietary fiber on the technological quality of pasta. *Food Sci. Technol. Int.* **2011**, *17*, 213–221. [CrossRef]
7. Abdel-Moemin, A.R. Analysis of phenolic acids and anthocyanins of pasta-like product enriched with date kernels (*Phoenix dactylifera L.*) and purple carrots (*Daucus carota L.* sp. *sativus var. atrorubens*). *J. Food Meas. Charact.* **2016**, *10*, 507–519. [CrossRef]
8. Crizel, T.M.; Araujo, R.R.; Rios, A.O.; Rech, R.; Flôres, S.H. Orange fiber as a novel fat replacer in lemon ice cream. *Food Sci. Technol.* **2014**, *34*, 332–340. [CrossRef]
9. Padalino, L.; Mastromatteo, M.; Lecce, L.; Spinelli, S.; Conto, F.; Del Nobile, M.A. Chemical composition, sensory and cooking quality evaluation of durum wheat spaghetti enriched with pea flour. *Int. J. Food Sci. Technol.* **2014**, *49*, 1544–1556. [CrossRef]
10. Gull, A.; Prasad, K.; Kumar, P. Effect of millet flours and carrot pomace on cooking qualities, color and texture of developed pasta. *LWT-Food Sci. Technol.* **2015**, *63*, 470–474. [CrossRef]
11. Ovando-Martinez, S.; Sa´yago-Ayerdi, E.; Agama-Acevedo, I.; Goni, L.A.; Bello, P. Unripe banana flour as an ingredient to increase the undigestible carbohydrates of pasta. *Food Chem.* **2009**, *113*, 121–126. [CrossRef]
12. Aguedo, M.; Kohnen, S.; Rabetafika, N.; Vanden-Bossche, S.; Sterckx, J.; Blecker, C. Composition of byproducts from cooked fruit processing and potential use in food products. *J. Food Compos. Anal.* **2012**, *27*, 61–69. [CrossRef]
13. Rabetafika, H.N.; Bchir, B.; Aguedo, M.; Paquot, M.; Blecker, C. Effects of processing on the composition and physicochemical properties of fibre concentrate from cooked fruit pomaces. *Food Bioprocess Technol.* **2014**, *7*, 749–760. [CrossRef]
14. Bchir, B.; Rabetafika, H.N.; Paquot, M.; Blecker, C. Effect of pear, apple and date fibres from cooked fruit by-products on dough performance and bread quality. *Food Bioprocess Technol.* **2014**, *7*, 1114–1127. [CrossRef]

15. Bchir, B.; Felfoul, I.; Bouaziz, M.A.; Gharred, T.; Yaich, H.; Noumi, E.; Snoussi, M.; Bejaoui, H.; Kenzali, Y.; Blecker, C.; et al Investigation of physicochemical, nutritional, textural, and sensory properties of yoghurt fortified with fresh and dried Spirulina (*Arthrospira platensis*). *Int. Food Res. J.* **2019**, *26*, 1565–1576.
16. Bouacida, S.; Amal, B.A.; Hayet, B.; Koubaier, B.; Christophe, B.; Nabiha, B. Chemical composition, cooking quality, texture and consumer acceptance of pasta with Eruca vesicaria leaves. *Int. J. Food Sci. Technol.* **2017**, *52*, 2248–2255. [CrossRef]
17. *Codex Alimentarius Hungaricus*; Directive 2-231: Dried Pasta Products; Ministry of Rural Development: Budapest, Hungary, 2013
18. AOAC International. *Official Methods of Analysis of AOAC International*, 16th ed.; Helrich, K., Ed.; Method Chapter 32 Cereal Foods; AOAC International: Gaithersburg, MD, USA, 1997.
19. Ainsa, A.; Honrado, A.; Marquina, P.L.; Roncalés, P.; Beltrán, J.A.; Calanche M., J.B. Innovative Development of Pasta with the Addition of Fish By-Products from Two Species. *Foods* **2021**, *10*, 1889. [CrossRef]
20. Mestres, C.; Colonna, P.; Bule, O. Characteristics of starch networks within rice flour noodles and mungbean starch vermicelli. *J. Food Sci.* **1988**, *53*, 1809–1812. [CrossRef]
21. Sozer, N.; Dalgıç, A.C.; Kaya, A. Thermal, textural and cooking properties of spaghetti enriched with resistant starch. *J. Food Eng.* **2007**, *81*, 476–484. [CrossRef]
22. Bchir, B.; Besbes, S.; Karoui, R.; Attia, H.; Paquot, M.; Blecker, C. Effect of air-drying conditions on physico-chemical properties of osmotically pre-treated pomegranate seeds. *Food Bioprocess Technol.* **2012**, *5*, 1840–1852. [CrossRef]
23. American Association of Cereal Chemistry. *Approved Methods of the American Association of Cereal Chemists*, 10th ed.; American Association of Cereal Chemists (AACC): St. Paul, MN, USA, 2000.
24. Chillo, S.; Laverse, J.; Falcone, P.M.; Protopapa, A.; Del Nobile, M.A. Influence of the addition of buckwheat flour and durum wheat bran on spaghetti quality. *J. Cereal Sci.* **2008**, *47*, 144–152. [CrossRef]
25. Aravind, N.; Sissons, M.J.; Fellows, C.M.; Blazek, J.; Gilbert, E.P. Effect of inulin soluble dietary fibre addition on technological, sensory, and structural properties of durum wheat spaghetti. *Food Chem.* **2012**, *132*, 993–1002. [CrossRef]
26. Zhen, M.; Joyce, I.B. Advances in the Design and Production of Reduced-Fat and Reduced-Cholesterol Salad Dressing and Mayonnaise: A Review. *Food Bioprocess Technol.* **2013**, *6*, 648–670.
27. Shreenithee, R.; Prabhasankar, P. Effect of different shapes on the quality, microstructure, sensory. and nutritional characteristics of yellow pea flour incorporated pasta. *Food Meas. Charact.* **2013**, *7*, 166–176. [CrossRef]
28. Kuchtová, V.; Kohajdová, Z.; Karovičová, J.; Mešterová, E. Use of pumpkin fiber for the preparation of pasta. *Chem. Listy* **2016**, *11*, 808–811.
29. Petitot, M.; Boyer, L.; Minier, C.; Micard, V. Fortification of pasta with split pea and faba bean flours: Pasta processing and quality evaluation. *Food Res. Int.* **2010**, *43*, 634–641. [CrossRef]
30. Rakhesh, N.; Fellows, C.M.; Sissons, M. Evaluation of the technological and sensory properties of durum wheat spaghetti enriched with different dietary fibres. *J. Sci. Food Agric.* **2015**, *95*, 2–11. [CrossRef]
31. Li, P.; Lu, W.; Hsieh, C.; Li, T.; Huang, D. Rheological Properties of Dough and Quality of Salted Noodles Supplemented with Djulis (*Chenopodium formosanum Koidz.*) Flour. *J. Agric. Sci.* **2015**, *6*, 84–92. [CrossRef]
32. Rosa-Sibakov, N.; Heinio, R.; Cassan, D.; Holopainen-Mantila, U.; Micard, V.; Lantto, R.; Sozer, N. Effect of bioprocessing and fractionation on the structural, textural and sensory properties of gluten-free faba bean pasta. *LWT-Food Sci. Technol.* **2016**, *67*, 27–36. [CrossRef]
33. Susanna, S.; Prabhasankar, P. A study on development of gluten free pasta and its biochemical and immunological validation. *LWT-Food Sci. Technol.* **2013**, *50*, 613–621. [CrossRef]
34. Wang, L.; Wei, D.; Haifeng, Q.; Hui, Z.; Qi, X. Effect of rice bran fibre on the quality of rice pasta. *Int. J. Food Sci. Technol.* **2018**, *53*, 81–87. [CrossRef]
35. Hoseney, C. *Principles of Cereal Science and Technology*; American Association of Cereal Chemists: St. Paul, MN, USA, 1999; pp. 269–274.
36. Kaur, N.; Gupta, A.K. Applications of inulin and oligofructose in health and nutrition. *J. Biosci.* **2002**, *27*, 703–714. [CrossRef]
37. Silva, M.; Spencer, J.; Shu-Hong, Y.; Martin, S.; Bonny, B.; Yao, O. Effect of selected dietary fibre sources and addition levels on physical and cooking quality attributes of fibre-enhanced pasta. *Food Qual. Saf.* **2019**, *3*, 117–127.
38. Wojtowicz, A.; Moscicki, L. Influence of legume type and addition level on quality characteristics, texture and microstructure of enriched precooked pasta. *LWT-Food Sci. Technol.* **2014**, *59*, 1175–1185. [CrossRef]
39. Islas-Rubio, A.R.; Calderon de la Barca, A.M.; Cabrera-Chavez, F.; Cota-Gastelum, A.G.; Beta, T. Effect of semolina replacement with a raw: Popped amaranth flour blend on cooking quality and texture of pasta. *LWT-Food Sci. Technol.* **2014**, *57*, 217–222. [CrossRef]
40. Sudha, M.L.; Rajeswari, G.; Venkateswara-Rao, O. Effect of wheat and oat brans on the dough rheological and quality characteristics of instant vermicelli. *J. Texture Stud.* **2011**, *43*, 195–202. [CrossRef]
41. Brennan, C.S.; Kuri, V.; Tudorica, C.M. Inulin-enriched pasta: Effects on textural properties and starch degradation. *Food Chem.* **2004**, *86*, 189–193. [CrossRef]
42. Hager, A.S.; Czerny, M.; Bez, J.; Zannini, E.; Arendt, E.K. Starch properties, in vitro digestibility and sensory evaluation of fresh egg pasta produced from oat, teff and wheat flour. *J. Cereal Sci.* **2013**, *58*, 156–163. [CrossRef]
43. Cemin, R.P.; Rios, A.O.; Thys, R.C.S. Use of broccoli (*Brassica oleracea italica*) leaves powder to produce fresh pasta. In Proceedings of the 20th Brazilian Congress of Chemical Engineering, Florinopolis, SC, USA, 19–22 October 2014; pp. 1–8.

44. Wojciech, M.; Maciej, T. Color difference Delta E—A survey. *Mach. Graph. Vis.* **2011**, *20*, 383–411.
45. Goesaert, H.; Brijs, K.; Veraverbeke, W.S.; Courtin, C.M.; Gebruers, K.; Delcour, J.A. Wheat flour constituents: How they impact bread quality, and how to impact their functionality. *Trends Food Sci. Technol.* **2005**, *16*, 12–30. [CrossRef]
46. Sun-Waterhouse, D.; Wadhwa, S.S. Industry relevant approaches for minimizing the bitterness of bioactive compounds in functional foods: A review. *Food Bioprocess Technol.* **2013**, *6*, 607–627. [CrossRef]
47. Lauková, M.; Kohajdová, Z.; Karovičová, J. Effect of hydrated apple powder on dough rheology and cookies quality. *Potravinarstvo* **2016**, *10*, 506–511. [CrossRef]
48. Aramouni, F.; Mahmoud, A. Physicochemical and sensory characteristics of no-bake wheat-soy snack bars. *J. Sci. Food Agric.* **2011**, *91*, 44–51. [CrossRef]
49. Martín-Esparza, M.; Raigon, M.; Raga, A.; Albors, A. Functional, Thermal and rheological properties of high fibre fresh pasta: Effect of tiger nut flour and xanthan gum addition. *Food Bioprocess Technol.* **2018**, *11*, 2131–2141. [CrossRef]
50. Sant'Anna, V.; Christiano, F.D.P.; Marczak, L.D.F.; Tessaro, I.C.; Thys, R.C.S. The effect of the incorporation of grape marc powder in fettuccini pasta properties. *LWT-Food Sci. Technol.* **2014**, *58*, 497–501. [CrossRef]

MDPI
St. Alban-Anlage 66
4052 Basel
Switzerland
Tel. +41 61 683 77 34
Fax +41 61 302 89 18
www.mdpi.com

Foods Editorial Office
E-mail: foods@mdpi.com
www.mdpi.com/journal/foods

www.ingramcontent.com/pod-product-compliance
Lightning Source LLC
Chambersburg PA
CBHW062036220926
51850CB00010B/2565
* 9 7 8 3 0 3 6 5 6 7 4 8 8 *